T0326261

Introduction to Transfer Phenomena
in PEM Fuel Cells

Series Editor
Alain Dollet

Introduction to Transfer Phenomena in PEM Fuel Cells

Bilal Abderezzak

First published 2018 in Great Britain and the United States by ISTE Press Ltd and Elsevier Ltd

ISTE Press Ltd
27-37 St George's Road
London SW19 4EU
UK

www.iste.co.uk

Elsevier Ltd
The Boulevard, Langford Lane
Kidlington, Oxford, OX5 1GB
UK

www.elsevier.com

Notices

Knowledge and best practice in this field are constantly changing. As new research and experience broaden our understanding, changes in research methods, professional practices, or medical treatment may become necessary.

Practitioners and researchers must always rely on their own experience and knowledge in evaluating and using any information, methods, compounds, or experiments described herein. In using such information or methods they should be mindful of their own safety and the safety of others, including parties for whom they have a professional responsibility.

To the fullest extent of the law, neither the Publisher nor the authors, contributors, or editors, assume any liability for any injury and/or damage to persons or property as a matter of products liability, negligence or otherwise, or from any use or operation of any methods, products, instructions, or ideas contained in the material herein.

For information on all our publications visit our website at http://store.elsevier.com/

British Library Cataloguing-in-Publication Data
A CIP record for this book is available from the British Library
Library of Congress Cataloging in Publication Data
A catalog record for this book is available from the Library of Congress
ISBN 978-1-78548-291-5

Printed and bound in the UK and US

Contents

Preface

The recognition of new sources of energy that are green and renewable is a necessity for young audiences in the scientific and technical field. These non-polluting energies contribute to a more protected environment against dangerous emissions, such as greenhouse emissions or those that affect air quality.

The operating principles of energy conversion devices that convert these renewable sources into useful energy must be known and controlled.

This educational book develops a broad overview of the different physical phenomena that take place within a fuel cell. This book is intended for students and young researchers in technical fields. It is essentially composed of five sections as follows.

Chapter 1 introduces hydrogen as an energy vector that can be produced in different ways and used in many applications. The different fuel cell technologies are presented in this chapter. A special interest in the proton-exchange membrane fuel cell is presented at the end of this chapter. In addition, a quiz specific to this introduction is provided as a good summary of the principles of hydrogen technology and PEM fuel cells.

In Chapter 2, charge transfer phenomena are discussed. This chapter covers the thermodynamic and chemistry aspects of a fuel cell, the flow rates of the reactants and products as well as some electrochemical notions. At the end of the chapter, polarization phenomena and an overview of charge transfer models are described.

Chapter 3 presents the mass transfer phenomena in the different layers of a fuel cell and, at different scales, in the membrane.

In Chapter 4, heat transfer phenomena are discussed. A broad overview of the energy balance equations is given for the overall fuel cell and also for each of its layers. The identification of heat sources and the thermal management of the fuel cell are presented at the end of this chapter.

At the end of each chapter, a conclusion will summarize the key concepts.

At the end of the book, a rich collection of bibliographic references can be found. Indeed, this includes the work of doctoral dissertations, the best French and English-speaking books that the author has been able to translate, masters dissertations and articles of scientific research.

All comments and suggestions are welcome, and it will be taken into consideration for a potential improvement of this book in subsequent editions.

Bilal ABDEREZZAK

June 2018

1

Introduction to Hydrogen Technology

Global issues such as pollution and global warming, as well as the increasing scarcity and depletion of fossil fuels, have highlighted the urgency of resorting to green and renewable energy sources. Moreover, the intermittency and irregularity of these sources have exposed the need to store energy in a chemical form, for example, as hydrogen.

Hydrogen appears to be an interesting environmentally friendly alternative to the oil or fuel that is currently used to produce energy. It is an energy vector that can store and transport energy.

Hydrogen can be stored and/or transported in different forms. The main method currently used to transport hydrogen between production and use sites is to use liquid hydrogen flowing through pipelines. The most discussed storage method currently concerns the use of hydrogen in embedded applications.

In addition to its non-polluting character, it allows the production of thermal and mechanical energy and electricity. However, the conversion of hydrogen into energy is the subject of several studies and research. In this initial part, we will present the hydrogen energy sector, with particular focus on the technology of the PEM fuel cell [AFH 18, AUP 13, BOU 03, GAR 00, GUP 08, MIS 13, RAJ 08, SAI 04].

1.1. Hydrogen as an energy vector

Hydrogen is the lightest and most abundant element in the universe. Mixed with oxygen, it can burn by releasing energy. It has a large amount of energy per unit mass; however, it contains a small amount of energy per unit volume at room temperature and at atmospheric pressure.

Hydrogen is an energy vector virtually non-existent in nature at the molecular level. This is why it must be produced by electrolysis, reforming of vapors or natural gas, gasification of biomass, or by oxidation and reforming of hydrocarbons or biomass. These methods are determined and controlled before hydrogen is used or stored. Nearly 95% of the hydrogen production is therefore derived from fossil fuels such as natural gas, oil or even coal (see Figure 1.1). The majority is produced from natural gas (48%) and it is used by industry for its chemical properties, particularly in ammonia plants (50% of global consumption) and in petroleum refineries (desulfurization of gasoline and diesel, production of methanol, etc.) [CON 18a].

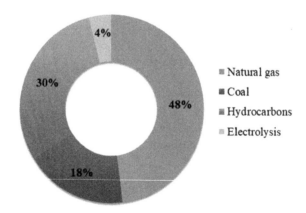

Figure 1.1. *Main sources of hydrogen production [RAJ 08]*

However, these processes do not help reduce our dependence on fossil fuels. Compared with other fuels, hydrogen has a higher calorific value (see Figure 1.2).

For further details, Table 1.1 shows a comparison of other elements [GAR 00, MIS 13].

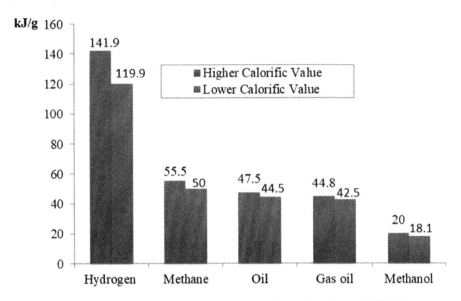

Figure 1.2. *Comparison of HCV and LCV for various fuels [AUP 13]*

Properties	Hydrogen	Methane	Methanol	Ethanol	Propane	Gasoline
Molecular mass (g/mol)	2.016	16.043	32.04	46.06	44.1	~107
Volumic mass (kg/m^3) at 20°C	0.08375	0.6682	791	789	1.865	751
Boiling point (°C)	−252.8	−161.5	64	78.5	−42.1	27-225
Flash point (°C)	−253	−188	11	13	−104	−43
Flammability limits in air (% volume)	4.0–75.0	5.0–15.0	6.7–36.0	3.3–19	2.1–10.1	1.0–7.6
CO_2 produced per unit energy	0	1	1.5			180
Auto-ignition temperature in air (°C)	585	540	385	423	490	230–480
Higher calorific value (MJ/kg)	142	55.5	22.9	29.8	50.2	47.3
Lower calorific value (MJ/kg)	120	50	20.1	27	46.3	44

Table 1.1. *Fuel comparison*

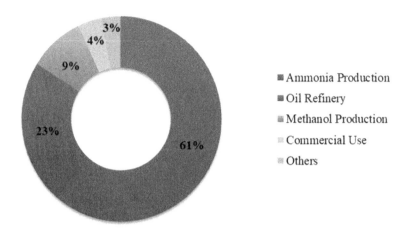

Figure 1.3. *Distribution of H₂ consumption*

The process is chosen according to many parameters (type of primary energy, purity, flows, etc.); it also depends on the target sector that intends to use this energy vector. Global consumption by application type is shown in Figure 1.3.

Today, less than 4% of the total production capacity of hydrogen is provided by electrolysis. This method is used only if the electricity is either inevitable (for renewable sources such as wind or photovoltaic), or cheap and/or if a high purity of the hydrogen is required. The increasing use of renewable sources is leading to the development of electrolysis, an attractive process for the development of these new energies [AFH 18].

There are even ways of producing hydrogen by wind energy, essentially by the electrolysis of water using a process that preserves the environment. Although the exploitation of wind energy and the electrolysis of water have been widely used in the past, their combination has not been widely applied at a commercial level. This production remains an efficient and safe way to store wind energy, especially in times of low energy demand and strong wind.

Hydrogen technologies applied to wind systems are still in the research and development stage. They are confined to small-scale applications. This energy sector can be considered as one of the most competitive energy markets once it is used as a primary energy for various mobile applications. The coupling of wind energy and the production of hydrogen are considered as a means of energy storage with several advantages.

First, using hydrogen as an energy vector while taking into account safety aspects is already understood, thanks to the numerous applications in chemistry. Hydrogen is also well suited for seasonal energy storage without energy loss over time. Still, water electrolyzers are able to process power fluctuations due to the intermittent nature of wind energy. Finally, wind–hydrogen systems have the potential for high-density energy storage with low operating and maintenance costs [GUP 08].

In Germany, the installation of the Prenzlau hybrid power plant is a real example to follow; it will not be long before the citizens become familiar with the expression "I go to the pump, to fill up the wind, 50 liters ...". A utopia? No, this is the first industrial hybrid plant in the world.

This project was supported by the German Chancellor Angela Merkel who laid the first stone on 21 April 2009 in Prenzlau near Berlin (see Figure 1.4).

Figure 1.4. *The electrolyzer of the Prenzlau power plant [MIS 13]*

On 25 October 2011, Brandenburg Minister-President Matthias Platzeck inaugurated this first hybrid plant of its kind combining wind, hydrogen and biogas in the presence of representatives of partner companies Enertrag AG, Total Deutschland GmbH, Vattenfall and Deutsche Bahn.

This facility is able to transform wind energy into hydrogen as a storage medium and energy vector! The hydrogen production capacity of the electrolyzer installed is approximately $120 \ \mathrm{Nm^3/h}$ of hydrogen and $60 \ \mathrm{Nm^3/h}$ of oxygen, with a purity of hydrogen reaching 99.997%. The pressure of the flue gas is about 15–20 mbar.

A compression system is used to achieve the pressure of 42 bar, and a set of five pressure vessels with a total capacity of 1,350 kg of hydrogen at the same compressor outlet pressure [MIS 13].

This technology is an important advance for the energy sector, since it allows the storage of wind power and its injection into the grid, according to need (see Figure 1.5).

Since the end of 2011, energy in the form of hydrogen has been stored and used at the Berlin–Brandenburg International Airport power station.

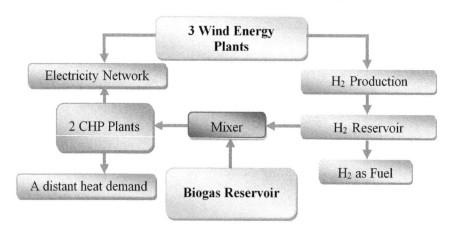

Figure 1.5. *Schematic of the Prenzlau hybrid power plant [MIS 13]*

This hybrid plant, which combines wind power, biogas and hydrogen, is part of the overall policy of developing renewable energies, supply of carbon-free energy, and better integration of intermittent renewable energies into the electricity grid.

It presents a perfectly balanced system between the fluctuation in the production of the different associated renewable energies and the actual electricity needs. Three wind turbines produce electricity and they are used, in part, to produce hydrogen. This CO_2-free vector is stored and, in addition to the biogas, can then be converted into electricity and heat in the event of peak consumption. Hydrogen also offers a CO_2-free flexible solution at total power stations in Berlin and Hamburg. The investment is estimated at approximately 21 million euros [MIS 13].

1.1.1. *Production methods*

Hydrogen is a good choice as a future non-polluting energy source for many reasons. Here are some of these reasons:

– hydrogen can be produced using several sources. It is quite renewable because the most abundant and clean starting element for producing hydrogen is water;

– the hydrogen can be stored in the gaseous state, in the liquid state or in the solid state. It can also be stored in different chemical substances such as methanol, ethanol or metal hydrides;

– the hydrogen can be produced by means of an electrochemical converter or efficiently converted into electricity;

– hydrogen can be transported and stored as safely as the fuels used today.

The hydrogen sector, or in other words, the production and applications of this very important energy vector, is composed of several variants. Figure 1.6 shows an overview of this sector. We can see three areas of application. Indeed, in the field of transport, hydrogen can be used in internal combustion engines and in electric motors powered by fuel cells.

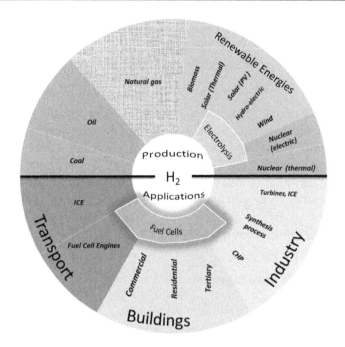

Figure 1.6. *Hydrogen technology sector*

The building sector also benefits from hydrogen, as there are hydrogen fuel cells suitable for commercial, residential, or even tertiary uses.

In industry, hydrogen can serve as the energy vector of several applications such as cogeneration by hydrogen fuel cells, turbines, or even internal combustion engines; its application in synthetic processes is also possible.

Hydrogen can therefore provide energy for all sectors of the economy: industry, homes, transport and portable devices. It can replace petroleum-based fuels for vehicles and it is an interesting source of electricity for communities. One of the main advantages of hydrogen, compared to fossil fuels, is its impact on the environment. Its consumption (by combustion or by electrochemical conversion) produces no pollutant and no carbon emissions.

1.1.2. *Storage technologies*

To store hydrogen, it is necessary to design tanks that are compact, lightweight, safe and inexpensive. This is decisive since it is precisely this storage option that makes hydrogen particularly attractive compared to electricity [CEA 18].

1.1.2.1. *Storage of liquids*

Storing hydrogen in liquid form is an attractive solution, and also in this form it is used in the space sector. However, hydrogen is, after helium, the most difficult gas to liquefy. This solution leads to high-energy expenditure and high costs, making it more difficult for the general public to use.

1.1.2.2. *Storage of high-pressure gas*

The storage of hydrogen in gaseous form is a promising option. However, the constraints are numerous. Light and bulky, the hydrogen must be compressed to the maximum to reduce the size of the tanks. Progress has been made: from 200 bar (pressure of industrial cylinders), the pressure has risen to 350 bar today, and the developments now concern tanks that can withstand pressures of 700 bar. This compression comes at a cost; even compressed at 700 bar, 4.6 liters of hydrogen are still needed to produce as much energy as with 1 liter of gasoline. The risk of hydrogen leakage must also be taken into consideration with regard to the flammable and explosive nature of this gas under certain conditions. Due to its small molecule size, hydrogen is able to pass through many materials, including some metals. This weakens some, making them brittle. The study of high-pressure storage therefore consists of essentially testing the resistance of materials to hydrogen under pressure. These materials must be resistant but relatively light (transportability requirements). The metal tanks currently used are still expensive and heavy compared to the amount of gas they can carry. Reservoirs, not metallic but made of polymeric materials, are being developed to meet these constraints.

1.1.2.3. *Low-pressure storage*

An alternative to using gas pressure tanks would be to store hydrogen in certain carbonaceous materials or metal alloys capable of absorbing hydrogen and releasing it when necessary. This storage mode is currently the subject of many studies.

1.1.3. *Distribution networks and associated risks*

Currently, the industrial distribution process used is to produce hydrogen in centralized units, then use it on site or transport it by pipelines.

Pipeline gas transportation connects major production sources at the main points of use. Hydrogen distribution networks via pipelines already exist in various countries to supply the chemical and petrochemical industries. Implementing these industrial infrastructures shows that one has a good control of the generation and the transport of hydrogen.

One caveat however: the transport cost is about 50% more expensive than that of natural gas and a unit of volume of hydrogen carries three times less energy than a unit volume of natural gas.

To distribute hydrogen, refueling infrastructures will have to be developed. The development of service stations does not appear to pose any particular technical problems. A total of 40 pilot stations already exist in the world, mainly in the United States, Japan, Germany and Iceland. However, it will take time for these service stations to cover the entire territory that requires this fuel, which could slow down its development in transport. To overcome this difficulty, some car manufacturers are considering using fuels that contain hydrogen rather than hydrogen alone. In this case, the reforming step takes place on board the vehicle. This process then becomes less interesting since this reforming produces carbon dioxide, the main cause of the greenhouse effect [CEA 18].

The accepted idea is that "hydrogen is more dangerous than traditional fuels", this assumption needs challenging. Of course, like any fuel, hydrogen can ignite and/or explode in case of leakage. Nevertheless, to compare the danger levels of traditional fuels and hydrogen, the nature of the assessed risk should be discerned [CON 18b].

1.1.3.1. *Risk of leakage*

The small size of the hydrogen molecule allows it to escape through extremely small openings. The risk of leakage is therefore higher with hydrogen than with other fuels.

1.1.3.2. *Flammability risk*

When a hydrogen layer is formed, the risk of flammability is significantly higher than for a slick of gasoline or natural gas. The energy required to ignite it is about 10 times lower than for natural gas.

1.1.3.3. *Risk of formation of an explosive layer*

Hydrogen disperses faster than traditional fuels. It is diluted four times faster in air than natural gas and 12 times faster than gasoline vapors. This volatility is a protective factor, limiting the formation of hydrogen layers.

Remember that hydrogen is used in industry. Prevention methods make it possible to limit the risks, such as the addition of an inert gas (such as CO_2) to reduce the flammability of hydrogen.

The use of a ventilation system and the deliberate ignition of hydrogen also prevent the formation of an explosive layer. If regulation is applied in an industrial environment, it still has to be defined for hydrogen consumer applications.

Safety measures would guarantee its use, particularly in the case of hydrogen vehicles and distribution infrastructure.

1.1.4. *Advantages and challenges to raise*

Here are some advantages and some challenges to raise, which are considered as downfalls for the use of hydrogen as an energy vector [AUP 13].

1.1.4.1. *Advantages*

There are six advantages:

– high energy molecule (120 MJ/kg vs. 50 MJ/kg for methane);

– neither pollutant nor toxic;

– its combustion in air generates only water;

– it is the lightest gas (wide diffusion in air = safety);

– it is easy to transport (pipes);

– depending on its mode of production: reduction of emissions, in particular CO_2.

1.1.4.2. *Disadvantages*

There are also six disadvantages:

– not present in the natural state;

– its lightness implies that its energy density is less favorable for storage and transport in gaseous form compared to natural gas ($10.8 \, MJ/m^3$ vs. $39.8 \, MJ/m^3$ for methane);

– flammability and detonation limit with air wider than for natural gas (4.1% – 72.5 vol. % and 5.1% – 13.5 vol. % for methane);

– minimum energy to be supplied to ignite it (10 times lower than that of conventional hydrocarbons);

– flame almost invisible;

– low public support.

1.2. Types of fuel cell

The development of the hydrogen industry is largely based on fuel cell technology. Its principle is not new, since it was discovered in 1839 by William R. Grove, who invented the first fuel cell. At the time, this English lawyer, an amateur researcher in electrochemistry, found that by recombining hydrogen and oxygen, it is possible to simultaneously create water, heat and electricity, and this is how the fuel cell was born.

In 1953, the engineer Francis T. Bacon realized the first industrial prototype of notable power (approximately 6 kW) [NGU 10]. Only NASA exploited this technology in the 1960s to provide electricity to some of its vessels such as Gemini and Apollo. Although the principle of the fuel cell seems simple, its implementation is complex and expensive, which hitherto prohibited its dissemination to the general public. Today, progress has been made and the potential applications are numerous, from the supercondensile fuel cell that produces only a few watts needed to power a mobile phone, to the battery capable of producing 1 MW to provide electricity to a building. The battery can also be used for embedded applications in the transport sector; there is now a whole range of fuel cells. The operating principle is always the same, but different technologies are in development [CEA 18]. Fuel cells will be one of the essential elements of the future hydrogen economy; they have the capacity to meet all our energy needs while offering high efficiency and very low-pollution technology.

1.2.1. *The different fuel cell technologies*

Several types of fuel cells are currently being researched. The classification of fuel cells is generally according to the nature of the electrolyte because it determines, on the one hand, the temperature at which the battery operates and, on the other hand, the ion type ensuring the ionic conduction [BLU 07]. For example:

– Polymer-Electrolyte Membrane Fuel Cell or Proton-Exchange Membrane Fuel Cell ~80°C (PEMFC);

– Direct Methanol Fuel Cell ~60–100°C (DMFC);

– Alkaline Fuel Cell ~100°C (AFC);

– Phosphoric Acid Fuel Cell ~200°C (PAFC);

– Molten Carbonate Fuel Cell ~700°C (MCFC);

– Solide Oxyde Fuel Cell ~800–1,000°C (SOFC).

Details on the operation of these different fuel cells are given in the following sections. Table 1.2 shows the type of electrolyte and the fields of application of fuel cells [BAR 05].

The proton-exchange membrane fuel cell (PEMFC) operates at 80°C with a polymer electrolyte. This is the most promising for transport.

Cell type	AFC	PEMFC	DMFC	PAFC	MCFC	SOFC
Electrolyte	KOH Solution	Proton-conducting polymer membrane	Proton-conducting polymer membrane	Phosphoric acid	Li_2CO_3, KCO_3 melted in $LiALO_2$	ZrO_2 and Y_2O_3
Ions in the electrolyte	OH^-	H^+	H^+	H^+	CO_3^{2-}	O^{2-}
Temperature	60–80°°C	60–100°C	60–100°C	180–220°C	600–660°C	700–1,000°C
Fuel	H_2	H_2 (pure or reformed)	Methanol	H_2 (pure or reformed)	H_2 (pure or reformed)	H_2 (pure or reformed)
Application domains	Spatial	Automobile, portable, cogeneration and maritime	Portable	Cogeneration	Cogeneration centralized electricity production, maritime	

Table 1.2. *Comparison of fuel cells*

1.2.1.1. *Operating principle of fuel cells*

The operating principle of fuel cells varies. It depends on the reagents (H_2, methanol, etc.), the nature of the electrolyte (solid or liquid), the residues of the electrochemical reaction and the operating temperature. The following sections provide an illustration of the operating mode for each type of battery.

1.2.1.2. *Polymer-electrolyte membrane fuel cell (PEMFC)*

The polymer-electrolyte membrane fuel cell (also called proton-exchange membrane (PEM)) provides high power density and low weight, reasonable cost and low volume. A PEM fuel cell comprises a negatively charged electrode (anode), a positively charged electrode (cathode and an electrolyte (the membrane). Hydrogen is introduced at the anode and oxygen at the cathode. The protons are transported from the anode to the cathode through the electrolyte membrane, and the electrons flow through an external circuit represented by the charge. A typical PEM fuel cell has the following reactions [SPI 07]:

– anode:

$$H_2(g) \longrightarrow 2H^+(aq) + 2e^- \hspace{3cm} [1.1]$$

– cathode:

$$\frac{1}{2}O_2(g) + 2H^+(aq) + 2e^- \longrightarrow H_2O \hspace{2cm} [1.2]$$

– overall reaction:

$$H_2(g) + \frac{1}{2}O_2(g) \longrightarrow H_2O(liq) + \text{Electricity} + \text{Heat} \hspace{1cm} [1.3]$$

These batteries work at a temperature below 100°C for a yield of around 50%. The low operating temperature gives them the ability to start up relatively quickly. These batteries are developed to power small- and medium-sized vehicles and fixed installations.

The power range of PEMFCs ranges from a few tens of watts to around 10 megawatts [BLU 07].

1.2.1.2.1. Operating characteristics

Table 1.3 summarizes the characteristics and performance of the PEMFC [BOU 07].

Characteristics and performance	
Operating temperature	60 to 80°C
Operating pressure	1 to 3 bar
Electrical efficiency	40 to 50%
Actual voltage	0.6 to 0.95 V
Current density	Up to A/cm²
Start-up time	Practically immediate
Response time to a rapid variation in demand	Very fast
Lifetime	1,000 to 2,000 hours (value in 2005) Up to 4,000 hours [HFC Consulting Group] Current tests to reach 8000 hours

Table 1.3. *Characteristics and performance of a PEMFC*

1.2.1.2.2. Advantages and disadvantages

These cells have some advantages [BLU 07, BOU 07]:

– very fast start-up time;

– very fast response time;

– compactness;

– insensitive to CO_2;

– the electrolyte is solid. There is therefore no risk of electrolyte leakage;

– the operating temperature is low;

– the mass power is high, the power density can reach 1 kW/kg.

However, there are some disadvantages:

– high membrane cost;

– the membranes must be permanently maintained in a good state of hydration to promote the transport of protons. Otherwise, there is a risk of damage to the membrane;

– they are sensitive to carbon monoxide which poisons the catalytic sites.

1.2.1.2.3. Aging

In a PEMFC cell, components are subject to chemical or mechanical aging, the main causes of which are:

– degradation of the membrane under the effect of temperature;

– loss of catalytic activity;

– heterogeneity of the materials used;

– moisture of the membrane not perfectly controlled.

1.2.1.3. *Alkaline fuel cell (AFC)*

Alkaline fuel cells (AFCs) have been used by NASA during space missions; they can achieve power generation efficiencies of up to 70%. The operating temperature of these batteries varies between 150 and 200°C. An aqueous alkali potassium hydroxide solution embedded in a matrix serves as an electrolyte. This is an advantageous configuration because the reaction at the cathode is fast in an alkaline electrolyte, which allows a better performance. Several manufacturers are examining ways to reduce the cost and improve the operational flexibility of these fuel cells. Alkaline fuel cells typically have powers of 300 watts to 5 kW [SPI 07]. The chemical reactions that take place in an alkaline fuel cell are:

– anode:

$$2H_2(g) + 4OH^-(aq) \longrightarrow 4H_2O(liq) + 4e^- \qquad [1.4]$$

– cathode:

$$O_2(g) + 2H_2O(liq) + 4e^- \longrightarrow 4OH^-(aq) \qquad [1.5]$$

– overall reaction:

$$2H_2(g) + O_2(g) \longrightarrow 2H_2O(liq) \qquad\qquad [1.6]$$

Figure 1.7 shows the operating principle of an AFC.

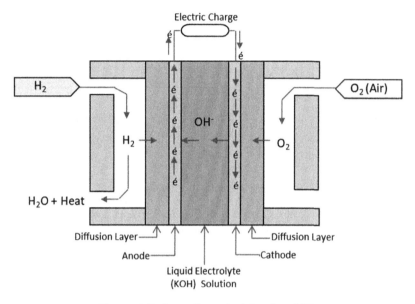

Figure 1.7. *Operating principle of an AFC*

Near atmospheric pressure, their operating temperature is limited to 80°C–90°C for a yield of 50%. However, this temperature can be higher (250°C) in the case of operating under pressure and with a highly concentrated electrolyte [BLU 07].

There are few applications for these batteries, the most significant ones being those of NASA with the Apollo program. The AFC batteries are still used in the shuttle (three by shuttles) to supply electricity at a power of between 2 kW and 12 kW (the maximum power is 16 kW) and a voltage of between 28 V and 32 V [WAG 03]. We can also mention Siemens who developed some batteries, including a 20 kW in the 1970s and 1980s, for submarines. The power range of AFC batteries is between 1 kW and 100 kW.

1.2.1.3.1. Operating characteristics

Table 1.4 summarizes the characteristics and performance of AFC [BOU 07].

Characteristics and performance	
Operating temperature	60 to 90°C
Operating pressure	1 to 5 bars
Electrical efficiency	> 60% (with pure hydrogen)
Actual voltage	0.7 to 1.0 V
Current density	100–200 m A/cm²
Start-up time	Several minutes
Response time to a rapid variation in demand	Relatively quick
Lifetime	5,000 hours (value in 2006)

Table 1.4. *Characteristics and performances of an AFC*

1.2.1.3.2. Advantages and disadvantages

Alkaline fuel cells have the advantage of allowing the use of non-precious catalysts based on nickel for the anode and activated carbon for the cathode.

They may also have the following advantages [BOU 07]:

– operation at atmospheric pressure and at low temperature;

– low costs of the electrolyte and catalysts;

– response time and fast start time;

– high electrical efficiency;

– operates at low temperature (below 0°C).

The main disadvantage posed by this type of battery is their sensitivity to carbon dioxide (CO_2) which, remember, is present in the air. This sensitivity implies a thorough purification of hydrogen (total elimination of CO_2) when it is obtained by reforming a hydrocarbon fuel. For this reason, the AFC battery has been abandoned for transport applications. Another reason for this is due to the liquid and corrosive state of the electrolyte which, subject to the transport constraints (vibrations, accelerations, etc.), can leak from the battery.

1.2.1.3.3. Aging

In the AFC cell, and in a closed cycle (without circulation), the electrolyte is diluted following the incomplete removal of the formed water. The electrolyte used is corrosive and it can attack the components with which it is in contact.

1.2.1.4. *Phosphoric acid fuel cell (PAFC)*

The phosphoric acid fuel cell (PAFC) is one of the few commercially available batteries. Hundreds of these batteries have been installed around the world. Most of these provide power ranging from 50 to 200 KW, and also 1 MW and 5 MW power plants have been built. The largest plant installed to date provides 11 MW of alternating current (AC), corresponding to a distribution network [ROS 99].

PAFCs have power generation efficiencies of 40%. Their operating temperature is between 150 and 300°C. PAFCs are poor ion conductors at low temperatures, and carbon monoxide (CO) tends to severely poison platinum in the catalyst [HIR 98]. The chemical reactions that occur in PAFCs are:

– anode:

$$H_2(g) \longrightarrow 2H^+(aq) + 2e^-$$ [1.7]

– cathode:

$$\frac{1}{2}O_2(g) + 2H^+(aq) + 2e^- \longrightarrow H_2O(liq)$$ [1.8]

– overall reaction:

$$\frac{1}{2}O_2(g) + H_2(g) + CO_2 \longrightarrow H_2O(liq) + CO_2$$ [1.9]

Figure 1.8 shows the operating principle of a PAFC.

The operating temperature is between 180°C and 210°C. At low temperatures, the electrolyte is not a good conductor and solidifies around 40°C. PAFC battery technology is the most mature in terms of development and commercialization. Indeed, stationary installations, up to 50 MW, have been set up. About 200 test facilities have been in operation worldwide, the main application of these batteries is cogeneration; the power range of PAFC batteries is between 200 kW and 50 MW [BLU 07].

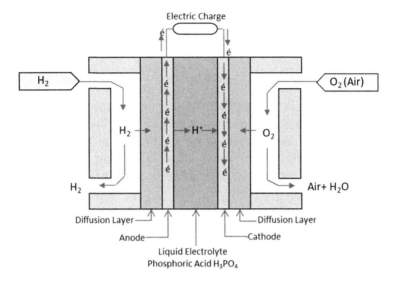

Figure 1.8. *Operating principle of a PAFC*

1.2.1.4.1. Operational characteristics

Table 1.5 summarizes the operational characteristics and performance of a PAFC [BOU 07].

Characteristics and performance	
Operating temperature	$\sim 200°C$
Operating pressure	Atmospheric pressure or up to 8 bars
Electrical efficiency	40 to 55%
Actual voltage	0.6 to 0.8 V
Current density	Up to 800 m A/cm²
Start-up time	1 to 3 hours
Response time to a rapid variation in demand	Very fast
Lifetime	> 40,000 hours (value in 2005)

Table 1.5. *Characteristics and performance of a PAFC*

1.2.1.4.2. Advantages and disadvantages

Compared with polymer membrane cells, PAFCs have the following advantages [BLU 07, BOU 07, SPI 07]:

– a better tolerance to carbon monoxide (tolerates ~1%) is the advantage of phosphoric acid batteries;

– not sensitive to CO_2;

– possibility of cogeneration (electricity-heat).

However, these batteries have a number of disadvantages:

– the electrolyte solidifies around 40°C, so the battery must be kept at a higher temperature to avoid damaging the electrodes;

– the strong corrosivity of the acid causes deterioration of the electrodes;

– sulfur sensitivity;

– relatively high start-up time;

– high cost of the catalyst (platinum).

1.2.1.4.3. Aging

Given the operating temperature of the PAFC cell, certain components are subject to modifications, such as the electrolyte, for example, which tends to degrade or evaporate; the catalyst also loses its catalytic activity during aging.

1.2.1.5. *Molten carbonate fuel cell (MCFC)*

In this type of cell, the reactions involved are more complex than in proton-exchange membrane, alkaline or phosphoric acid fuel cells. The ionic conduction is ensured by the migration of carbonate ions (CO_3^{2-}) from the anode to the cathode through the electrolyte constituted by molten carbonates. One of the special features of MCFC batteries (such as the SOFC battery) is that the operating temperature allows the use of carbon monoxide (CO) as a fuel. We must remember that carbon monoxide is considered a poison for low and medium temperature batteries. This can come from the process of reforming a hydrocarbon. There can therefore be two kinds of electrochemical reactions at the anode. The first is the main reaction of converting the energy contained in hydrogen into electricity. The second is a reaction that can take place if carbon monoxide is present in the fuel (it therefore does not occur if the battery is supplied with pure hydrogen), it is favorable to the battery since it also makes it possible to produce electricity [BLU 07, BOU 07, SPI 07]:

– anode:

$$H_2(g) + CO_3^{2-} \longrightarrow H_2O(g) + CO_2(g) + 2e^-$$ [1.10]

– cathode:

$$\frac{1}{2}O_2(g) + CO_2(g) + 2e^- \longrightarrow CO_3^{2-}$$ [1.11]

– overall reaction:

$$H_2(g) + \frac{1}{2}O_2(g) + CO_2(g) \longrightarrow H_2O(g) + CO_2(g)$$ [1.12]

Figure 1.9 shows the operating principle of a MCFC.

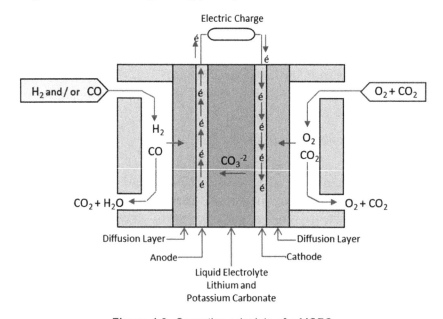

Figure 1.9. *Operating principle of a MCFC*

The operating temperature is between 600°C and 700°C. At this temperature, the carbonates are melted and are maintained by capillarity in the matrix.

MCFCs are mainly used in stationary applications. An example is a 2 MW natural gas mini-plant in the United States that has been operating close to 4,000 hours. The power range is between 500 kW and 10 MW.

1.2.1.5.1. Operating characteristics

Table 1.6 summarizes the characteristics and performance of the MCFC [BOU 07].

Characteristics and performance	
Operating temperature	650 to 700°C
Operating pressure	One to several bars
Electrical efficiency	55%
Actual voltage	0.75 to 0.90 V
Current density	Up to 200 mA/cm^2
Start-up time	Up to several hours
Response time to a rapid variation in demand	Dependent on size
Lifetime	Several thousand hours

Table 1.6. *Characteristics and performance of a MCFC*

1.2.1.5.2. Advantages and disadvantages

The high operating temperature of these batteries gives them some advantages:

– they have a better tolerance to compounds that contaminate hydrogen, particularly carbon monoxide (CO), which can be used as fuel along with hydrogen;

– the reforming of hydrocarbons (methane, propane, etc.) or of methanol can be carried out inside the cell itself;

– high efficiency;

– nickel catalyst;

– possibility of cogeneration.

However, one of the difficulties encountered with this type of battery relates to the corrosion of nickel oxide by the electrolyte; other disadvantages include:

– controlling the CO_2 taken from the anode and reinjected at the cathode complicates the operation;

– the relatively long start-up time;

– sensitivity to sulfur;

– the dissolution of the cathode (nickel);

– low current density.

1.2.1.5.3. Aging

In a MCFC, adverse reactions can occur. This is due to the high operating temperatures as well as the corrosive nature of the electrolyte.

Indeed, the dissolution of Ni^{2+} ions in the electrolyte at the cathode and their diffusion towards the anode (with the possibility of short circuit if they are reduced in the form of nickel).

In addition, the mechanical stability of the electrodes is also affected at these temperatures, resulting in deformations detrimental to high efficiency. The structure of the $liAlO_2$ electrolyte matrix can also be modified (increase in particle size and porosity).

1.2.1.6. *Solid oxide fuel cell (SOFC)*

Solid oxide fuel cells (SOFCs) appear to be used in large, high power plants such as power plants in industry.

A solid oxide system is generally made of a hard ceramic material, comprising zirconium solid oxide and a small amount of yttria to replace the liquid electrolyte. The operating temperature of these solid oxide batteries can reach 1,000°C. Their efficiency can reach 60 to 85% with cogeneration, and when the output power exceeds 100 KW [SPI 07]. Internal reactions are as follows:

– anode:

$$H_2(g) + O_2^- \longrightarrow H_2O(g) + 2e^- \tag{1.13}$$

– cathode:

$$\frac{1}{2}O_2(g) + 2e^- \longrightarrow O_2^- \tag{1.14}$$

– overall reaction:

$$H_2(g) + \frac{1}{2}O_2(g) \longrightarrow H_2O(g) \tag{1.15}$$

Figure 1.10 shows the operating principle of a SOFC.

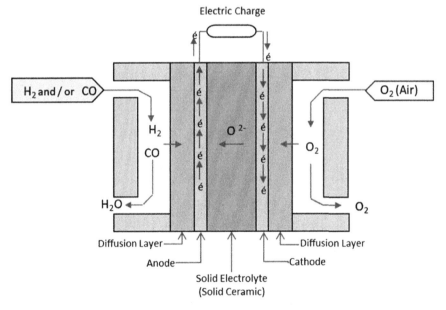

Figure 1.10. *Operating principle of a SOFC*

The operating temperature is between 800°C and 1,000°C. At this temperature, the ceramic has a very good ionic conductivity.

SOFCs are mainly found in stationary applications; however, some car manufacturers see these fuel cells being used in the future because they accept a wide variety of fuels. The power range of SOFC batteries is between 1 kW and 10 MW [BLU 07].

1.2.1.6.1. Operating characteristics

Table 1.7 summarizes the operating characteristics of a SOFC [BOU 07].

Characteristics and performance	
Operating temperature	900 to 1,000°C
Operating pressure	1 to 10 bars
Electrical efficiency	60%
Actual voltage	0.7 to 1.15 V
Current density	Up to 1,000 m A/cm²
Start-up time	Up to several hours (depending on size)
Response time to a rapid variation in demand	Slow
Lifetime	> 30,000 hours (value in 2006)

Table 1.7. *Characteristics and performance of a SOFC*

1.2.1.6.2. Advantages and disadvantages

The high temperature has many advantages:

− it allows for cogeneration;

− it makes possible the internal reforming of hydrocarbons such as methane (CH_4); in the long term, the direct oxidation of methane is envisaged;

− the stability of the electrolyte;

− low cost nickel catalysts.

However, a temperature close to 1,000°C has certain disadvantages:

– the use of specific materials, resistant to this temperature, is vital;

– corrosion is rapid;

– very long start-up time;

– sensitivity to the variation in operating temperature;

– evacuation of heat.

1.2.1.7. Direct methanol fuel cell (DMFC)

The great potential for portable uses of fuel cells has generated enormous interest for a fuel cell that can run directly on methanol. The direct methanol fuel cell (DMFC) uses the same polymer membrane as the PEM fuel cell. However, the fuel used in DMFC is methanol instead of hydrogen. Methanol flows through the anode as fuel and it breaks down into protons, electrons, water and carbon dioxide. The advantages of methanol are: it is available everywhere, and it can easily be reformed from gasoline or biomass. Although its energy density is only one-fifth of that of hydrogen per unit weight, since it is in liquid form, it offers more than four times the amount of energy per volume compared to hydrogen under a pressure of 250 atmospheres [BRU 99, HIR 98]. The chemical reactions that occur in this fuel cell are:

– anode:

$$CH_3OH(liq) + H_2O(liq) \longrightarrow CO_2 + 6H^+ + 6e^- \qquad [1.16]$$

– cathode:

$$6H^+ + \frac{3}{2}O_2 + 6e^- \longrightarrow 3H_2O(liq) \qquad [1.17]$$

– overall reaction:

$$CH_3OH(liq) + \frac{3}{2}O_2(g) \longrightarrow CO_2(g) + 2H_2O(liq) \qquad [1.18]$$

Figure 1.11 shows the operating principle of a DMFC.

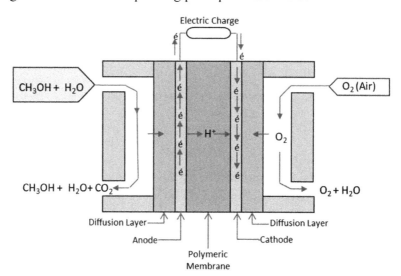

Figure 1.11. *Operating principle of a DMFC*

1.2.1.7.1. Operating characteristics

Table 1.8 summarizes the characteristics and performance of a DMFC [BOU 07].

Characteristics and performance	
Operating temperature	~ 60°C
Operating pressure	~ 1 to 3 bars
Electrical efficiency	~ 30 to 50%
Actual voltage	0.4 to 0.7 V
Current density	100 to 200 mA/cm²
Start-up time	Practically immediate
Response time to a rapid variation in demand	Very fast

Table 1.8. *Characteristics and performance of a DMFC*

1.2.1.7.2. Advantages and disadvantages

The migration of methanol from the anode to the cathode through the membrane results in poisoning of the catalyst on the oxygen side. To solve this problem, it will be necessary to design membranes less permeable to methanol than those currently used.

In addition, the low oxidation temperature of methanol requires a large amount of catalyst compared to the PEMFC cell. However, it has the advantage of using a liquid fuel and requires no reformer.

Another disadvantage of these batteries is the toxicity of methanol, which is why some companies are exploring another way, that is, the direct ethanol fuel cell (DEFC), in which methanol is replaced by ethanol. Currently, DMFC batteries are more efficient than DEFCs.

1.2.1.7.3. Aging

Chemical or mechanical aging may occur in DMFC batteries and the main causes are:

– degradation under the effect of temperature;

– loss of catalytic activity (poisoning of the catalyst, agglomeration of the particles, etc.);

– heterogeneity of the materials used;

– insufficiently controlled humidity.

1.2.1.8. *Reversible fuel cell (RFC)*

The RFC is a closed-loop system that can provide electricity, heat and water. Hydrogen is provided by the electrolysis of water using renewable energy such as solar, wind and geothermal.

This battery is used by NASA in the Helios project, a drone equipped with a fuel cell and solar panels: the solar panels allow the drone to fly by day and store hydrogen so that the fuel cell can operate during the night.

1.2.1.9. *Metal–air fuel cell*

The most common fuel cells in this category are zinc–air batteries, although aluminum–air or even magnesium–air batteries are also commercially available. In all cases, the operations (reactions) that take place are all the same and the specialists have got into the habit of calling them Zinc Fuel Cells or Zinc–Air Fuel Cells (ZAFC) [LAR 03].

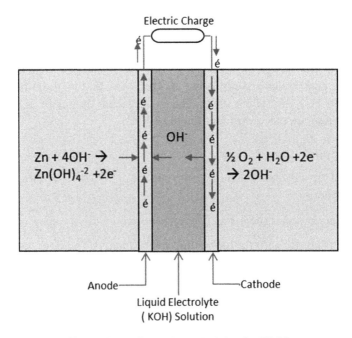

Figure 1.12. *Operating principle of a ZAFC*

The ZAFC shares some of the characteristics of other fuel cells as well as batteries. The electrolyte of a ZAFC cell is a solid ceramic material and the ionic conductivity is ensured by the transfer of ions (OH⁻) (see Figure 1.12). The ZAFC operates at a temperature of 700°C, which allows it to reform a large number of hydrocarbons inside the fuel cell, thus avoiding the use of an external reformer. This type of fuel cell has a zinc tank that regenerates the fuel. Indeed, zinc is in the form of small granules that will react (they will be consumed) with the electrode (the anode) by releasing electrons, and thus forming an electric current; at the cathode, oxygen reacts with the electrons from the anode, forming potassium zincates K_2ZNO_2 [SME 00]. The efficiency of the system using zinc regeneration is approximately 30 to 50%.

The advantage of this technology (zinc–air) is the remarkably high specific energy rate. In addition, the cost is relatively low since zinc is an abundant element.

The disadvantage of this type of battery is that the zinc anode is consumed by an electrochemical reaction and it must be replaced regularly. The applications of these batteries are generally dedicated to electric vehicles, and consumables of electronic devices for military applications.

1.2.1.10. *Proton ceramic fuel cell (PCFC)*

These are the new types of fuel cells based on the use of a ceramic electrolyte with high proton conductivity at high temperature and high energy efficiency [COO 03]. This type of battery is fundamentally different from other fuel cells, because the transport of protons through the electrolyte becomes more efficient at high temperatures. PCFC batteries have the same kinetic and thermal characteristics as an MCFC or SOFC because of their high operating temperature ($\sim 700°C$) [SPI 07].

1.2.1.11. *Bio-fuel cell (BFC) or microbial fuel cell (MFC)*

The bio-fuel cell is an electrochemical device that converts biochemical energy into electrical energy. Indeed, it is a redox reaction of an organic substrate (glucose, methanol, acetate, etc.) using micro-organisms or enzymes as catalysts. This type of cell is promising, but long term [LAR 03]. It should be noted that PCFC and BFC batteries are the most recent types of batteries and they are still in the research and development stage. The direct methanol (DMFC) or direct ethanol (DEFC) fuel cells directly consume the hydrogen contained in alcohol. Very compact, they are destined to power microelectronics and portable tools.

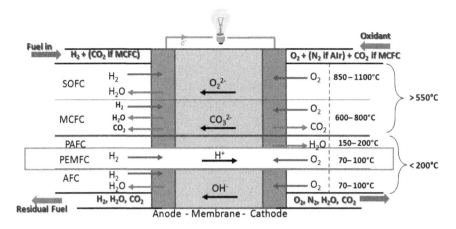

Figure 1.13. *Different types of fuel cells*

The solid oxide fuel cell (SOFC) is attractive for stationary applications because its very high operating temperature (around 800°C) makes it possible to use natural gas directly without reforming. In addition, the residual heat can be exploited in turn directly, or used to produce electricity using a gas turbine. In this case, the overall yield could reach 80%. Figure 1.13 summarizes the different types of fuel cells.

1.2.2. *Fuel cell and their applications*

Fuel cells would be a very useful accompaniment to the choice of available energy, because they provide electrical energy for uses currently limited in energy capacity. For example, one of the most monotonous things is laptops. Their limitation lies in their energy reserves since they have an autonomy of only 2 hours. Existing fuel cells today would give laptops an 8-hour battery life. After that, a quick refill of chemical fuel would allow the user to continue his work. Each sector requires fuel cells for different reasons, as described in the following sections. Fuel cells have multiple applications that are divided into several categories:

– automobile: vehicle, bus, etc.;

– stationary: generation of heat and electricity (residential, public building, e.g. swimming pool, hospital, school, apartments and emergency generator);

– portable: phone, laptop, camera, camping or military equipment;

– space: boat and submarine.

1.2.2.1. Stationary sector

Large fuel cells, for stationary use, provide sufficient power to power a building or offices.

These fuel cells can also produce additional electrical energy that can be injected into the grid and sold by the producer.

Stationary fuel cells are especially advantageous for isolated houses where the electricity grid is not available. In addition, fuel cells can be associated with a wind turbine or photovoltaic system to provide an efficient hybrid solution for the supply of electrical energy.

1.2.2.2. Transport market

Transport will benefit from fuel cells because fossil fuels will soon become scarce, which will inevitably lead to higher prices.

Legislation around the world is also becoming increasingly demanding in terms of carbon emissions and pollutants. Some countries enact laws requiring the reduction of emissions and imposing a certain proportion of "zero emission" clean vehicles.

Fuel cell-powered vehicles offer higher efficiencies than conventional vehicles powered by other fuels.

1.2.2.3. Portable sector

One of the main markets for fuel cells in the future will be the portable equipment sector. The military field also requires powerful and long-lasting power supplies for the troops. There are many portable devices on the market that will use fuel cells to gain autonomy. These devices include laptops, mobile phones, portable DVD players, video cameras, game stations, iPods, etc. A fuel cell will run a device as long as it is fueled. The trend in electronic devices, the convergence of functions and their limitations lie in the energy demand. This is why an energy source such as a fuel cell, which is able to provide more energy and for longer, allows the development of devices with multiple functions.

Table 1.9 summarizes the potential areas of use of fuel cells.

Application	Portable	Residential	Transport	Power station
Power	1–100 W	1–10 kW	10–100 kW	100 kW–10 MW
PEMFC				
DMFC				
PAFC				
AFC				
MCFC				
SOFC				

Table 1.9. *The potential domains of use of fuel cells*

1.2.3. *Advantages and issues to improve*

Fuel cells are often presented as the solution of the future in the electric power generation and automotive sectors. This is justified by their many advantages, for example, the fact that they benefit from a wide range of operating conditions. This means that some fuel cells can be used at room temperature, while others can operate at high temperatures when this is an advantage.

Fuel cell efficiency is determined by pressure, temperature and humidity during operation. The efficiency can often be improved (depending on the type of fuel cell) by raising the temperature, pressure, humidity or by optimizing certain variable parameters of the cell. The possibility of improving these variables depends on its area of use, because considerations of weight, size or cost can be an important decision factor.

Other advantages can be summarized as follows:

– high energy yields, even at partial charge: from 40 to 70% electrical, more than 85% in all (electricity and heat);

– low noise levels;

– low emissions (particularly, in terms of CO, NOx, CnHm and particulates, but depend on the fuel used and the type of application);

– they are of modular construction;

– various operating temperatures: this allows the use of heat by coupling with a turbine or for applications ranging from hot water to steam;

– no rotating parts.

Whereas the issues to overcome are:

– the cost: downfall of fuel cells, the price is much higher ensuring competitiveness. The key points are the catalyst (platinum), membranes, bipolar plates, periphery, etc.;

– weight and volume: especially if you want to integrate it into a vehicle;

– the lifetime: they must last more than 40,000 hours in stationary applications;

– thermal integration: from heat recovery to heat evacuation according to the fuel cell and the application.

1.3. The proton-exchange membrane fuel cell

Among the various types of fuel cells currently developed, the so-called "solid polymer" sector has been adopted by almost all automobile stakeholders in the world. Generally known as PEMFC, it is also of interest to manufacturers for stationary (hundreds of kW), portable (up to 100 W) and transportable (around 100 kW) applications. This technology is gaining attention for three main reasons: its relatively low operating temperature (< 100°C) allows for simplified technology to ensure a quick startup and easy evacuation of heat produced at room temperature, which is essential for automotive applications; it is also insensitive to the presence of CO_2 in the air, unlike the alkaline type; and finally, its all-solid technology can therefore allow significantly superior lifetimes to liquid electrolyte fuel cells, as well as a simpler industrialization. This makes it possible to envisage a future cost that is compatible with the targeted market, especially since it offers a significantly greater compactness than other types. This type of fuel cell provides a high power density and a lower weight, a reasonable cost and a low volume. A PEM fuel cell consists of a negatively charged electrode (anode), a positively charged electrode (cathode) and an electrolyte (the membrane). Hydrogen is introduced at the anode and oxygen is introduced at the cathode. The protons are transported from the anode to the cathode through the electrolyte membrane, and the electrons flow through an external circuit represented by the charge [SPI 07] (see Figure 1.14). These batteries work at a temperature below 100°C with efficiency of around 50%.

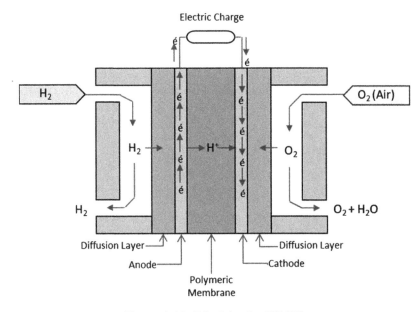

Figure 1.14. *Principle of a PEMFC*

The low operating temperature gives them an ability to start up relatively quickly. These fuel cells are developed to power small and medium-sized vehicles and fixed installations. The power range of PEMFC ranges from a few tens of watts to around 10 megawatts [BLU 07]. PEMFC technology is currently one of the largest research and development areas in the world. The main aim of the research is to increase performance (in terms of efficiency, lifetime, compactness and mass) and to reduce the cost (by a factor of 5 to 100 depending on the type of application).

On the contrary, it is important to note that a fuel cell module, to operate, must be associated with ancillary equipment that supplies reagents (hydrogen and air), their conditioning (pressure, hygrometry and flow rate) and the control of the reaction products (water, heat, electricity). These ancillaries can represent up to 60% in mass and volume of the system and an energy expenditure of around 20%. Fuel storage is the main technological hurdle, because it determines, for certain applications, the autonomy of the system.

Most car manufacturers have been developing fuel cell models over the last decade and they have produced at least one prototype. The best reason for developing fuel cell vehicles is their efficiency, as well as their very low polluting or zero emissions.

Fuel cell vehicles have at least one of these advantages:

– the fuel cell produces all the energy required to operate the vehicle;

– a battery can be used for start-up;

– a fuel cell performs best at constant power, therefore for accelerations or for power peaks, devices such as batteries or ultracapacitor batteries are used in addition;

– in certain vehicle models, the primary source of energy may be batteries, in which case the fuel cell becomes the secondary source whose task is to recharge the batteries;

– the fuel cell can supply energy to all or part of the vehicle.

The operating temperature of a car fuel cell varies from 50 to 80°C; a temperature above 100°C would improve heat transfer and simplify the cooling of the battery, but most vehicles use proton-exchange membrane fuel cell (PEMFC) or direct methanol fuel cell (DMFC), whose operating temperature cannot reach 100°C, to facilitate water circulation inside the cell [HOR 09].

1.3.1. *The basic structure of the PEMFC*

A PEMFC cell is an electrochemical device that converts chemical energy of a reaction directly into electrical energy while releasing heat and producing water.

It essentially consists of a membrane electrode assembly (MEA) with a thickness of 500 to 600 μm (see Figure 1.4). Electrons flow through a conductor if necessary. The anode and the cathode include a catalyst for producing electricity from the chemical reaction.

The chemical energy of the reactants is converted into electrical energy, heat and water by the catalyst (electrode with a mixture of carbon and platinum). The fuel (hydrogen) and the oxidant (oxygen) move to the catalyst layers where the chemical reaction occurs. It must be noted that the water and residual heat produced by the fuel cell must be constantly removed as these residues may present risks for the battery [HOR].

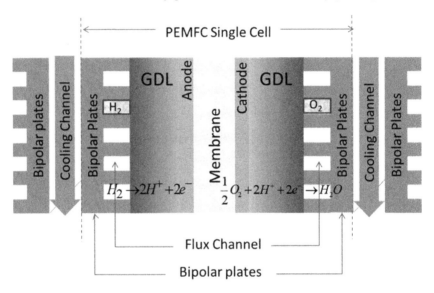

Figure 1.15. *Detailed diagram of a PEMFC*

The PEMFC cell consists of five main elements (see Figure 1.15). For its proper functioning, a PEMFC cell requires different auxiliary systems. The flow and pressure of the gas supply is controlled. Generally, with the hydrogen being stored at high pressure, system relaxation is necessary. On the contrary, if oxygen is drawn from the air, a compressor is essential to ensure continuous supply under pressure, this can have a rather significant energy cost.

Depending on the materials used in the MEA, the gases must be injected with a certain humidity; their humidification can be done actively or passively [CHU 09]. Finally, gas purification accessories are sometimes necessary, especially in the case of hydrogen produced from the reforming of hydrocarbons. The main components required for the proper functioning of a PEMFC cell are presented in the following sections.

1.3.1.1. *The electrolyte layer*

The electrolyte layer (electrolytic polymer membrane) is at the core of a fuel cell. It allows electrons to travel in the required manner by attracting protons and allowing them to pass through the membrane while maintaining their proton state. In fact, the membrane behaves like an acid solution in which there are negatively charged fixed sulfonic sites (SO^-_3). These sites dissociate water molecules to create H^+ protons. The membrane transports ions and water through diffusion and osmotic drive effects (these will be discussed in detail in Chapter 3 on mass transfer). The electrons flow through the external circuit to supply the charge in the form of an electric current. The hydrogen protons thus cross the electrolyte layer to join the cathode, and then recombine with the oxygen to form water. The membrane is a complex polymer that must have the following properties: it must be both a good ionic conductor, an electronic insulator, impervious to gas and stable mechanically and chemically.

1.3.1.1.1. Types of membranes

The membrane used in PEMFC is composed of a fluorocarbon skeleton (based on perfluorocarbon sulfonic acid ionomer (PSA) similar to Teflon (hydrophobic tetrafluoroethylene (TFE)) with hydrophilic sulfonic groups (SO^-_3)), and even several perfluorosulfonate monomers. These membranes are most commonly known as Nafion[TM] and is in the form of poly(perfluoro-sulfonylethoxy propylene vinyl ether) (PSEPVE). It was developed in the 1970s by DuPont de Nemours. It is generally characterized by its concentration in fixed charges: the higher it is, the better the conduction of protons. Its chemical structure is shown in Figure 1.16 [COL 08].

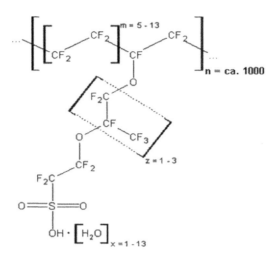

Figure 1.16. *Chemical structure of Nafion*

Other suppliers also provide polymer membranes: Dow Chemical (Dow Membrane with a shorter chain), Asahi Glass (Flemion), Asahi Chemical (Aciplex), Gore, Pemeas, etc. These polymers are physically in the form of sheets, rolls or in solution, which is cast to be shaped [BOU 07]. Nafion membranes are available in different sizes and thicknesses; they are usually marked with the letter "N" followed by three or four digits, the first two of which represent the equivalent weight (EW) expressed in $g.eq^{-1}$, divided by 100, whereas the remaining numbers represent the thickness of the membrane expressed in mills (1 mill = 1/1,000 inch = 25.4 μm). The typical thicknesses of Nafion^{TM} membranes are 2; 3.5; 5; 7 and 10 mills, that is, 50; 89; 127; 178 and 254 μm respectively. For example, Nafion^{TM} N117 has an equivalent weight of 1 100 g eq^{-1} and a thickness of 7 mills (178 μm) [BAR 05]. The smaller the thickness, the greater the membrane is permeable to gas, but the fewer the proton conductionlosses.

1.3.1.1.2. The concept of equivalent weight

The equivalent weight (EW) is the weight of polymer per mole of active sites (SO_3^-). It corresponds to the inverse of the ion exchange capacity (IEC) which represents the amount of sulfonate groups introduced per gram of polymer in a fuel cell [BAR 05]. It is expressed as follows:

$$EW = 100n + 446 \qquad\qquad [1.19]$$

with (n) the number of TFE groups per PSEPVE monomer [DOY 03], the optimum equivalent weight is generally between 900 and 1350 g eq^{-1}; a value of 1 100 g eq^{-1} is generally found [SPI 07] and [COL 08]. Nafion^{TM} membranes have good stability in an acid environment. They cannot function in temperatures above 100°C, and they need to be maintained in a state of sufficient hydration to ensure good ionic conduction. Other products are beginning to emerge to replace Nafion^{TM} membranes, but their performance must still be assessed.

1.3.1.1.3. The membrane water content

It should be noted that when these membranes are placed in a humid environment, they adsorb (or desorb) a certain amount of water and swell or deflate in all directions until reaching a state of equilibrium [ELL 01].

Hydration of the membrane is generally characterized by its water content (λ_m), but it is also defined as being the number of water molecules contained in the membrane per active site. The amount of water adsorbed is greater when the environment is wet, since the membrane adsorbs more water when it is immersed in liquid water than in a saturated humidity environment; this phenomenon is called the Schröder paradox [VAL 06].

1.3.1.1.4. The structural model of Nafion^{TM}

A structural model of hydrated Nafion membrane at room temperature was envisaged by Gierke *et al.* [GIE 81] and Hsu and Gierke [HSU 83] in the late 1970s. For a long time, it was the most widespread and a representation is given in Figure 1.17.

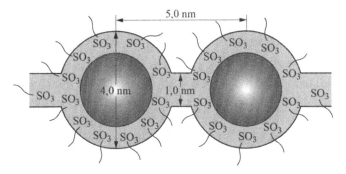

Figure 1.17. *Structural model of Nafion^{TM}*

Indeed, when the membrane is humidified, aggregates of solvated ionic sites called clusters, about 4 nanometers (nm) in diameter, are supposed to form in the matrix of the polymer.

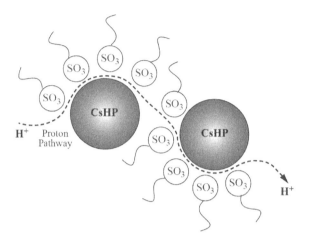

Figure 1.18. *The path of the proton in the membrane structure of Nafion*

Beyond a percolation threshold, these aggregates are interconnected by 1 nm channels for the diffusion of ions and water molecules within the membrane. The eventual pathway taken by the proton through the membrane is shown in Figure 1.18.

Other works on this same topic have emerged and they seem to agree on Nafion having a porous structure, whose pores, occupied on the surface, are a few nanometers in size.

1.3.1.2. *The catalytic layer*

The reaction of the fuel cell occurs in the catalyst layers (CLs) which consist of the anode and the cathode. At the anode, the hydrogen molecules are separated into electrons and protons. At the cathode, oxygen recombines with the protons of hydrogen to form water. These electrodes are therefore active layers and these are the sites of electrochemical half reactions. When the operating temperature remains low, typically $T < 100°C$, the reactions are slow and the addition of a catalyst is essential [LAR 03]. In PEMFCs, platinum remains the most commonly used catalyst, even if alternative solutions now exist [BAL 02].

The electrode typically consists of a mixture of an ionomer and suspended black platinum on coal particles (Pt/C); the charge catalyst is usually between 0.1 and 0.4 mg/cm^{-2} for both the anode and the cathode [MIK 07].

The reaction mechanisms in the active layers are relatively complex. These catalytic layers are often the thinnest of the fuel cell, with a thickness of 5 to 30 microns. They are also the most complex because they must contain several types of gas, water and facilitate chemical reactions. These catalytic layers are made of a porous carbon (granular porous materials), platinum or platinum/ruthenium material [MIK 01]. The reactions that occur in the catalytic layer are exothermic; it is therefore necessary to remove heat from the fuel cell. The heat can be removed by convection through the channels and conduction into the solid part of the catalytic layer, the gas diffusion layer and the bipolar plates. Since water is produced in PEMFC, condensation and evaporation of this water affects the heat transfer. This is why the condensation and evaporation of water and the temperature of the fuel cell are intimately linked [APP 88].

In all cases, these layers, also called "active" layers, must always be accessible to gas, but they must also be able to ensure the transfer of protons via the membrane and the transfer of electrons to the external electrical circuit from the anode to the cathode; the manufacture of electrodes must consider the best compromise to respect these rules of operation. Over time, there may be a decrease in catalyst activity, which, although not consumed during the reaction, gradually decompose and no longer provide contact with the electrolyte. In addition, the electrodes are very sensitive to carbon monoxide, which may be one of the products of the reforming of hydrogen. A few ppm of CO is sufficient to poison active sites, resulting in a decrease in potential [NGU 10, RAM 05].

Operation in pure oxygen or oxygen-enriched air substantially improves the performance of the fuel cell. The diffusion of oxygen through the diffusers and especially through the pores of the electrode is facilitated. However, this improvement should be considered alongside the additional costs and difficulties related to the complexity of the system [RAM 05].

1.3.1.3. *The gas diffusion layers*

The gas diffusion layers (GDL) are a fibrous porous medium that has two main functions: to ensure a uniform distribution of reactive gases on the surface of the electrodes, and the transport of electrons to or from the external electrical circuit. They are made of thin carbon fibers, a porous and hydrophobic material. The thickness of these fibers is between 250 and 400 µm, and the perforations or pores available are 4 to 50 µm in size [NGU 10]. These gas diffusion layers must also serve as a support for the electrolyte and its structure must facilitate the evacuation of water that would prevent the reaction from occurring.

1.3.1.4. *Bipolar plates*

Bipolar plates are made of graphite or metal; they evenly distribute the fuel and the oxidant to the cells of the fuel cell. They also collect the generated electric current at the output terminals.

In a single-cell fuel cell, there is no bipolar plate; however, there is a single-sided plate that provides the flow of electrons. In fuel cells that have more than one cell, there is at least one bipolar plate (flow control exists on both sides of the plate). Bipolar plates provide several functions in the fuel cell.

Some of these functions include the distribution of fuel and oxidant inside the cells, the separation of the different cells, the collection of the electric current produced, the evacuation of the water from each cell, the humidification of the gases and cooling of the cells. Bipolar plates also have channels which allow the passage of reactants (fuel and oxidant) on each side. They form the anode and cathode compartments on opposite sides of the bipolar plate. The design of the flow channels may vary; they may be linear, coiled, parallel, comb-like or evenly spaced as shown in Figure 1.19.

The materials are chosen based on chemical compatibility, corrosion resistance, cost, electrical conductivity, gas diffusion ability, impermeability, ease of machining, mechanical strength and their thermal conductivity.

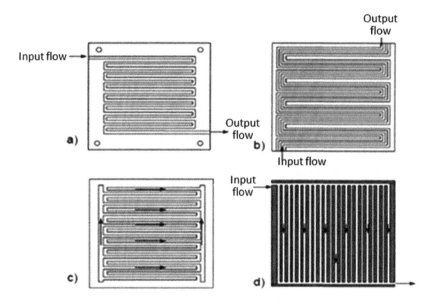

Figure 1.19. *The different types of bipolar plate [COL 08].*
a) Coiled flow channels; b) multiple coil flow channels;
c) parallel flow channels; d) interdigitated flow channels

1.3.2. *PEMFC design and configuration*

Since most electrical devices require voltage levels and power that cannot be produced by a single cell, in fuel cells, a number of cells are connected in series forming a "battery" [COL 08]. These cells, all-similar, are separated by plates allowing the flow of fluids and form what is known as a stack (see Figure 1.20).

Increasing the number of cells in the stack amplifies the voltage produced, while increasing the area of each cell increases the available current. In the traditional design of bipolar batteries, fuel cells have many cells (or elements) connected in series. The cathode of a cell is connected to the anode of the next cell.

MEAs, seals, bipolar plates and endplates are the typical layers of a fuel cell. These cells are assembled firmly.

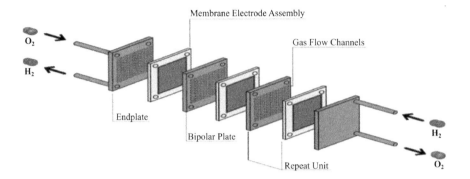

Figure 1.20. *Exploded view of a stacked PEM fuel cell*

The most common configuration is shown schematically in Figure 1.21. Each cell (MEA) is separated by a plate provided with flow channels to allow the distribution of the fuel and oxidant. Most fuel cells have this configuration, regardless of size, type or fuel used.

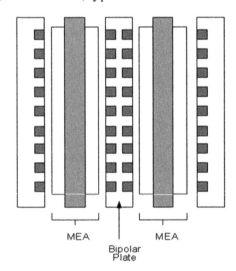

Figure 1.21. *The path of the proton in the membrane structure of Nafion*

The performance of fuel cells depends on the flow rate of the reagents and irregular distribution, which causes irregular operation of cells. The reactive gases must be distributed to all the cells by the same main channels.

1.3.3. *Operation and aging problem*

Table 1.10 summarizes the characteristics and performance of the PEMFC [BOU 07].

Characteristics and performance	
Operating temperature	60 to 80°C
Operating pressure	1 to 3 bars
Electrical efficiency	40 to 50%
Actual voltage	0.6 to 0.95 V
Current density	Up to A/cm^2
Lifetime	1,000 to 2,000 hours (value in 2005) Up to 4,000 hours [HFC Consulting Group] Current tests reach 8,000 hours (with a very quick start-up time)

Table 1.10. *The performances of a PEM fuel cell*

These fuel cells have some advantages [BLU 07, BOU 07]:

– very fast start-up time;

– very fast response time;

– compactness;

– insensitive to CO_2;

– the electrolyte is solid. There is therefore no risk of electrolyte leakage;

– the operating temperature is low;

– the mass power is high, the power density can reach 1 kW/kg.

However, they have some disadvantages:

– high membrane cost;

– the membranes must be maintained permanently in a good state of hydration to promote the transport of protons. Otherwise, there is a risk of damage to the membrane;

– they are sensitive to carbon monoxide, which poisons catalytic sites.

In a PEMFC cell, components are subject to chemical or mechanical aging, the main causes of which are:

– degradation of the membrane under the effect of temperature;

– loss of catalytic activity;

– heterogeneity of the materials used;

– moisture of the membrane not perfectly controlled.

1.3.4. *The fuel cell and its technical system entourage technique*

The proper functioning of the PEMFC requires different ancillary systems such as supply, humidification, cooling, control and regulation. The overall system (battery core and auxiliaries) is called "fuel cell system"; it is presented in Figure 1.22 [BLU 07, RAM 05].

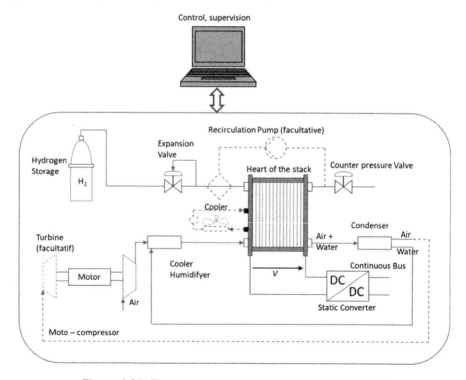

Figure 1.22. *The control and monitoring systems of a PEM*

1.3.4.1. *The supply system*

The flow and pressure of the gas circuits is controlled. These control systems make it possible to control the supply conditions of the fuel cell.

To improve performance, it is possible to work at high airflows in order to evacuate the water produced and to limit the effects of the low partial pressure of oxygen in the air. However, the use of compressors confers a high energy cost [RAM 05].

At the anode, there is little increase in the hydrogen stoichiometry and it is necessary to integrate a hydrogen recirculator to recover the unused hydrogen (output) and reinject it (input). The oxygen supplying the fuel cell is usually taken from the ambient air to prevent the storage of oxygen. However, the air only contains 21% of oxygen, and in order to increase the partial pressure of oxygen, it is sometimes necessary, for high power, to compress the air. The use of a compressor is also to the detriment of the overall efficiency of the system. It is powered by the battery and uses significant energy, accounting for approximately 20% of the electric power delivered by the battery [BLU 07]. Finally, there is the storage device (or production) of hydrogen. Tolerance of combustible gases to impurities is an important parameter. Some poisons such as sulfur or carbon monoxide can seriously damage the battery. In the case of prototype cells, the hydrogen used is usually a very pure gas [RAM 05].

1.3.4.2. *The control system*

As seen previously, the system has a large number of interconnected and interdependent ancillaries, in order to ensure the proper functioning of the system (overall efficiency, operational safety, taking into account constraints).

It is necessary to have an overall control system to control the different subsystems (static converter, gas flow, compressor control, water pump required for cooling, humidifier temperature, etc.).

1.3.4.3. *The static converter*

The static converter makes it possible to form the electrical energy produced by the fuel cell. The voltage of the battery, which is relatively low, is not constant as a function of the delivered current. The converters are of the following types: continuous DC, voltage booster (low voltage, high current) and unidirectional current. The static converter generally provides an interface between the battery and a continuous bus where, depending on the application, other energy storage components such as batteries, supercapacitors or other (DC/DC or DC/AC) converters are connected, which make it possible to form an interface between the DC bus and the charge (e.g. an electric motor) [BLU 07].

1.3.4.4. *The cooling circuit*

The conversion of hydrogen and oxygen into electricity also produces heat, which must be removed in order to keep the battery temperature constant. The evacuation of this heat, for low power, can be done using a fan (forced air convection) which can be the same as that supplying the cell with air. In the case of higher power batteries, forced air convection is no longer sufficient to remove heat. Other types of more complex cooling systems such as water cooling systems are used. This system operates in a closed loop thanks to a pump that circulates deionized water in a secondary circuit inside the cell [BLU 07].

1.3.4.5. *The humidification circuit*

Generally, PEMFC-type cells require a gas humidification circuit so that the membrane is not dehydrated (drying of the membrane results in an increase in ionic strength of the membrane) or the electrodes are not oversaturated due to excess water. This circuit therefore has the role in humidifying the incoming gas, usually from the water produced by the battery recovered via a capacitor. The control of the humidification is very delicate. More and more research aims to find solutions to overcome such a circuit by increasing, for example, the operating temperature of the battery (high temperature membranes) [BLU 07].

1.4. Conclusion

This first part allowed us to present the hydrogen sector, to define the functioning of a highly developed electrochemical system, the fuel cell. Special interest was later attributed to PEM battery technology. However, these benefits are still outweighed by the significant marketing costs.

In addition, there are still some technological barriers that prevent the reduction of these costs and the increase in the fuel cell lifetime.

An in-depth understanding of the transfer mechanisms governing the electrical performance of the cell makes it possible to optimize the operating conditions of these systems (supply and humidification of the gases, thermal and hydraulic management of the cell core, etc.).

1.5. Questions

1) Can hydrogen be found in its natural state?

2) What does the hydrogen sector mean?

3) How many methods exist to produce hydrogen?

4) What are the top three applications for hydrogen?

5) Can hydrogen be used in internal combustion engines?

6) What would be the best energy application of hydrogen?

7) Can we find a way to store hydrogen without compressing it at high pressure?

8) How many types of fuel cells are there? What are they?

9) What are the main features that distinguish different types of batteries?

10) Can we find batteries dedicated to certain applications?

11) What could you say about the life of a battery relative to its operating cycles?

12) How many elements form the structure of a cell in the PEM stack?

13) What is the role of each element of the PEM stack?

14) Could you describe the operating principle of a stack of cells in the PEM stack?

15) When drawing a diagram, try to place the PEM battery in its technical system and name each ancillary.

2

Charge Transfer Phenomena

2.1. Introduction

The movement of the charged species within the PEM cell occurs due to the creation of an electric current; the charges involved are known as protons and electrons. Their production at the anode and their recombination at the cathode are ensured by the catalyzed reactions of oxidation of hydrogen and the reduction of oxygen, respectively. These two half-reactions, located at the electrodes, allow the conversion of chemical energies into electrical energy. In this chapter, we will describe the transfer of charges within the PEM cell and its electrical performance.

First, we will give an overview of the theoretical characterization of the reversible electric power for such a system. The introduction of the thermodynamic variables H, S and G will make it possible to define respectively the total energy supplied to the system, the part dissipated as heat, and finally, the electric energy available. The variations in these variables with temperature and pressure are detailed.

Then, the real (non-reversible) operation of the fuel cell is studied. The different overvoltages in the PEM fuel cell are identified and quantified. For this, some electrochemical reminders are given to introduce the Nernst and Butler–Volmer equations.

Using the correlations of the membrane proton conductivity, it is possible to calculate the ohmic, activation and concentration overvoltages for a cell in operation.

Finally, the different models that we present allow the calculation of the operating voltage of a cell according to the supply conditions, the temperature and the desired current density [RAM 05].

2.2. Thermodynamics and chemistry of the PEM fuel cell

The PEM fuel cell is an electrochemical system involving electron transfer to the surface of the electrodes; the produced current is proportional to the surface of the electrode and it is expressed as current density. Understanding the phenomena occurring within the fuel cell requires knowledge of thermodynamics, chemistry, electrochemistry, hydraulics and electricity [HIR 94, LAR 03].

2.2.1. *The base reaction*

The electrochemical reaction, in the case of a fuel cell, occurs simultaneously at both sides of the membrane (anodeand cathode) as follows:

– anode:

$$H_2 \longrightarrow 2H^+ + 2e^- \qquad\qquad [2.1]$$

– cathode:

$$\frac{1}{2}O_2 + 2H^+ + 2e^- \longrightarrow H_2O \qquad\qquad [2.2]$$

– overall reaction:

$$H_2 + \frac{1}{2}O_2 \longrightarrow H_2O \qquad\qquad [2.3]$$

These reactions may have other intermediate steps giving rise to undesirable reactions [OHA 09].

2.2.2. *Heat reaction*

The overall reaction given by equation [2.3] is the same as the hydrogen combustion reaction, which is exothermic. The enthalpy of the chemical reaction is the difference between the heat (enthalpies) of formation of the products and reactants [BAR 05], which gives for the overall equation:

$$\Delta H = (h_f)_{H_2O} - (h_f)_{H_2} - \tfrac{1}{2}(h_f)_{O_2} \qquad [2.4]$$

The enthalpy of formation of an element, by convention, is equal to zero; however, the heat of formation of the liquid water (at 25°C) is -286 kJ.mol^{-1}. The negative sign means a release of heat and the equation [2.3] becomes:

$$H_2 + \frac{1}{2}O_2 \longrightarrow H_2O(\text{liq}) + 286 \text{ kJ.mol}^{-1} \qquad [2.5]$$

The sign of the value (286 kJ.mol^{-1}) becomes positive because it is on the side of the products of the electrochemical reaction. It should be noted that equation [2.5] is valid for temperature and pressure conditions: 25°C and atmospheric pressure, respectively. This amount of heat released by the electrochemical reaction, occurring in the cell, is also known as the calorific value and is otherwise defined by the amount of heat released by complete combustion of one mole of hydrogen [BAR 05, BLU 07].

2.2.3. Electrical work

In a fuel cell, and because there is no combustion, the notion of the Higher (HCV) or Lower (LCV) Calorific Value represents the maximum amount of thermal energy that could be extracted from hydrogen during the electrochemical reaction [BAR 05]. The question that arises: as we want to recover electrical energy after the reaction, can all the energy contained in hydrogen be converted by the battery into electrical energy? Obviously no! In any chemical (or electrochemical) reaction, there is a disorder that appears, in our case it is the creation of an entropy; in other words, it is the irreversible aspect of the reaction, and only a part of what was called HCV will be converted into electricity. This is the Gibbs free energy (G) or theoretical (maximum) electrical work of the fuel cell, in (kJ.mol^{-1}) [BOU 07], given by the following equation:

$$W_e = n \cdot F \cdot E = -\Delta G \qquad [2.6]$$

and:

$$\Delta G = \Delta H - T\Delta S \qquad\qquad [2.7]$$

where:

 $-$ n is the number of electrons exchanged in the elementary electrochemical reaction, in the case of the hydrogen/oxygen reaction, $n_e = 2$;

 $-$ F is the Faraday constant, electric charge of one mole of electrons, F = 96 485 $C.mol^{-1}$;

 $-$ E is the electromotive force (emf) of the fuel cell.

In the same way, like formation enthalpies in equation [2.4], the following equation represents the difference between entropy of products and reactants:

$$\Delta S = (S_f)_{H_2O} - (S_f)_{H_2} - \tfrac{1}{2}(S_f)_{O_2} \qquad\qquad [2.8]$$

Table 2.1 summarizes the enthalpy and formation entropy values for the products and reactants of the reaction produced in the fuel cell (at 25°C and at atmospheric pressure) [BAR 05].

	H_f (kJ.mol^{-1})	S_f (kJ.K^{-1}.mol^{-1})
Hydrogen, H_2	0	0.13066
Oxygen, O_2	0	0.20517
Liquid water, H_2O (l)	−286.02	0.06996
Water vapor, H_2O (g)	−241.98	0.1884

Table 2.1. *Enthalpies and entropies of formation*

Based on these data presented in Table 2.1, it can be deduced that only 237.34 kJ.mol^{-1} of all the thermal energy involved (286.02 kJ.mol^{-1}) can be converted into electricity and 48.68 kJ.mol^{-1} is transformed into heat.

2.2.4. *Empty voltage*

The theoretical potential of a fuel cell is the potential difference between the cathode and the anode, this voltage is denoted as (E). The superscript character (0) is used to designate the battery voltage under standard temperature and pressure conditions (25°C, 1 atm) [BLU 07]. It is said to be empty because it is calculable without the presence of an electric charge that consumes the current delivered by the fuel cell.

At equilibrium, the Faraday equation connecting the free energy to the electrode potential is written as:

$$\Delta G^0 = -n \cdot F \cdot E^0 \qquad [2.9]$$

where (F) is the Faraday constant and its value is 96 485 C.mol^{-1} and the number of electrons exchanged is always equal to 2, hence:

$$E^0 = \frac{-\Delta G^0}{2F} \qquad [2.10]$$

In the case where the water formed is in liquid form, the theoretical voltage of an H_2/O_2 fuel cell is:

$$E^0 = \frac{-\Delta G^0}{2F} = \frac{-237.34}{2 \times 96485} = 1.23\,V \qquad [2.11]$$

In the case where the water formed is in vapor form, the theoretical voltage of a H_2/O_2 fuel cell is:

$$E^0 = \frac{-\Delta G^0}{2F} = \frac{-228.6}{2 \times 96485} = 1.185\,V \qquad [2.12]$$

The difference is due to the change in the Gibbs free energy during water vaporization.

2.2.5. *Effect of pressure*

The preceding equations are valid for a pressure equal to the atmospheric pressure. However, the battery can operate at different pressures, ranging from atmospheric pressure to several bars [BAR 05, SPI 07]. For a process considered isothermal, the change of free energy of Gibbs can be written as:

$$dG = V_m \cdot dP \qquad\qquad [2.13]$$

where:

– V_m is the molar volume, in $[m^3.mol^{-1}]$;

– P is the pressure, in [Pa].

For an ideal gas, we write:

$$PV_m = RT \qquad\qquad [2.14]$$

Replacing equation [2.13] in [2.14], we find:

$$dG = RT\frac{dP}{P} \qquad\qquad [2.15]$$

Then, after integration:

$$G = G^0 + RT\ln\left(\frac{P}{P_0}\right) \qquad\qquad [2.16]$$

where (G^0) and (P_0) are respectively the Gibbs free energy and pressure at standard pressure and temperature. During a chemical reaction of the following type:

$$jA + kB \rightarrow mC + nD \qquad\qquad [2.17]$$

The change in Gibbs' free energy is a function of the products and reactants, so that:

$$\Delta G = mG_C + nG_D - jG_A - kG_B \qquad\qquad [2.18]$$

Following substitution, equation [2.16] becomes:

$$\Delta G = \Delta G^0 + RT \ln \left[\frac{\left(\dfrac{P_C}{P_0} \right)^m \left(\dfrac{P_D}{P_0} \right)^n}{\left(\dfrac{P_A}{P_0} \right)^j \left(\dfrac{P_B}{P_0} \right)^k} \right] \qquad [2.19]$$

In the case of a H_2/O_2 fuel cell, we write:

$$\Delta G = \Delta G^0 - RT \ln \left(\frac{P_{H_2O}}{P_{H_2} \cdot P_{O_2}^{0,5}} \right) \qquad [2.20]$$

The potential is expressed as follows:

$$E = E^0 - \frac{RT}{nF} \ln \left(\frac{P_{H_2} \cdot P_{O_2}^{0,5}}{P_{H_2O}} \right) \qquad [2.21]$$

This equation is known as the Nernst law (at equilibrium).

2.2.6. *Effect of temperature*

In an open circuit (in the case where the battery is not connected to an external electrical charge) [PER 08, SAI 06], for the anode, we write:

$$E_{theo\_anode} = \frac{\Delta G_{anode}}{nF} \qquad [2.22]$$

and for the cathode:

$$E_{theo\_cathode} = \frac{\Delta G_{cathode}}{nF} \qquad [2.23]$$

The electrical potential, that is, the difference between two electrode potentials, is written as:

$$E_{theo} = E_{theo\_cathode} - E_{theo\_anode} \qquad [2.24]$$

According to the Nernst law, we can write:

– for the anode:

$$\Delta G_{anode} = \Delta G^0_{anode} - RT \ln\left(P_{H_2}\right) \qquad [2.25]$$

– for the cathode:

$$\Delta G_{cathode} = \Delta G^0_{cathode} - \frac{RT}{2} \ln\left(P_{O_2}\right) \qquad [2.26]$$

The energy that is actually recoverable is that of the standard free enthalpy (Gibbs free energy standard), which is denoted as (ΔG^0):

– for the anode:

$$\Delta G^0_{anode} = \Delta H^0_{anode} - T\Delta S^0_{anode} \qquad [2.27]$$

– for the cathode:

$$\Delta G^0_{cathode} = \Delta H^0_{cathode} - T\Delta S^0_{cathode} \qquad [2.28]$$

The variation of enthalpy and entropy is a function of the temperature, $T_0 = 298.15$ K:

$$\Delta H^0(T) = \Delta H^0(T_0) + \int_{T_0}^{T} \Delta C_p.dT \qquad [2.29]$$

and for the entropy:

$$\Delta S^0(T) = \Delta S^0(T_0) + \int_{T_0}^{T} \frac{1}{T} \Delta C_p.dT \qquad [2.30]$$

The specific heat C_p is a function of temperature; the following equation shows an empirical relation usable in our case:

$$C_p = \alpha + \beta T + \gamma T^2 \qquad [2.31]$$

where T denotes the operating temperature of the cell; α, β and γ are the coefficients that depend on the gas. Table 2.2 gives the values of these coefficients [PER 08, SAI 06]:

Substance	α (J . mol^{-1} . K^{-1})	β (J . mol^{-1} . K^{-2})	γ (J . mol^{-1} . K^{-3})
Hydrogen, H_2	29.038	$-0.8356 . 10^{-3}$	$2.0097 . 10^{-6}$
Oxygen, O_2	25.699	$12.966 . 10^{-3}$	$-3.8581 . 10^{-6}$
Water, H_2O	30.33	$9.6056 . 10^{-3}$	$1.1829 . 10^{-6}$

Table 2.2. *Numerical values of empirical coefficients*

The following equations allow us to calculate the variation of the standard Gibbs free energy for temperatures other than 298.15 K. The index (i) designates (anode or cathode).

For the enthalpy:

$$\Delta H_i^0(T) = \Delta H_{298}^0 + \alpha(T - 298) + \beta \frac{T^2 - 298^2}{2} + \gamma \frac{T^3 - 298^3}{3} \qquad [2.32]$$

For the entropy:

$$\Delta S_i^0(T) = \Delta S_{298}^0 + \alpha \ln\left(\frac{T}{298}\right) + \beta(T - 298) + \gamma \frac{T^2 - 298^2}{3} \qquad [2.33]$$

Knowing that:

$$\Delta H_{reaction} = \Delta H_{products}^0 - \Delta H_{reac\,tan\,ts}^0 \qquad [2.34]$$

and:

$$\Delta S_{reaction} = \Delta S_{products}^0 - \Delta S_{reac\,tan\,ts}^0 \qquad [2.35]$$

which means we will have at the anode:

– for the enthalpy:

$$\Delta H_{anode} = 0 - \Delta H_{H_2}^0 \qquad\qquad [2.36]$$

– for the entropy:

$$\Delta S_{anode} = 0 - \Delta S_{H_2}^0 \qquad\qquad [2.37]$$

then, for the enthalpy at the cathode:

$$\Delta H_{cathode} = \Delta H_{H_2O}^0 - \frac{1}{2}\Delta H_{O_2}^0 \qquad\qquad [2.38]$$

and for the entropy at the cathode:

$$\Delta S_{cathode} = \Delta S_{H_2O}^0 - \frac{1}{2}\Delta S_{O_2}^0 \qquad\qquad [2.39]$$

For temperatures below 100°C, the variation in the value of (C_p), (ΔH) and (ΔS) is not very significant, but at high temperatures (in some types of fuel cell such as SOFC), this change should not be negligible.

The above equations and tables allow us to calculate the theoretical anode and cathode potentials for any given temperature [PER 08, SAI 06].

For the anode:

$$E_{theo\_anode} = \frac{T\Delta S_{H_2}^0 - \Delta H_{H_2}^0 - RT\ln\left(P_{H_2}\right)}{2F} \qquad\qquad [2.40]$$

and for the cathode:

$$E_{theo\_cathode} = \frac{\Delta H_{H_2O}^0 + \frac{1}{2}\Delta S_{O_2}^0 - \frac{1}{2}\Delta H_{O_2}^0 - T\Delta S_{H_2O}^0 - \frac{RT}{2}\ln\left(P_{O_2}\right)}{2F} \qquad\qquad [2.41]$$

2.2.7. *Theoretical efficiency*

The efficiency of any energy system is defined as the ratio of the energy collected (useful) to the energy supplied. In the case of the fuel cell (see Figure 2.1), the energy received is the electrical energy produced by the cell and the energy supplied to the system is the enthalpy of hydrogen, more specifically the HCV of hydrogen [BAR 05, OHA 09, SPI 07].

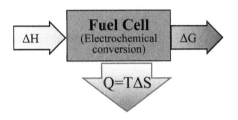

Figure 2.1. *Energy balance for a fuel cell*

The efficiency of the fuel cell is written as:

$$\eta\% = 100 \times \left[\frac{\text{Electrical Energy Produced}}{\text{Energy released by the reaction}} \right] = 100 \times \frac{\Delta G}{\Delta H} \qquad [2.42]$$

The efficiency can also be written in the form:

$$\eta_{\%} = 100 \times \left[1 - T\left(\frac{\Delta S}{\Delta H} \right) \right] \qquad [2.43]$$

The relationship defining the efficiency of a fuel cell in the case where the produced water is in the liquid state, the HCV value is used, it is written as:

$$\eta\% = 100 \times \frac{237.34}{286.2} \approx 83\% \qquad [2.44]$$

This value of 83% is theoretical. The LCV is often used to calculate the efficiency of the cell, because not only it has a high value but also it could be compared to that of a heat engine whose efficiency is always calculated by the LCV. In this context, the produced water is as a vapor; the yield becomes:

$$\eta\% = 100 \times \frac{228.74}{241.98} \approx 94.5\% \qquad [2.45]$$

For thermal engines, the efficiency is written as:

$$\eta_\% = 100 \times \left[1 - \sqrt{\frac{T_C}{T_H}} \right] \qquad [2.46]$$

Figure 2.2 shows the behavior of both yields with respect to temperature [BOU 07].

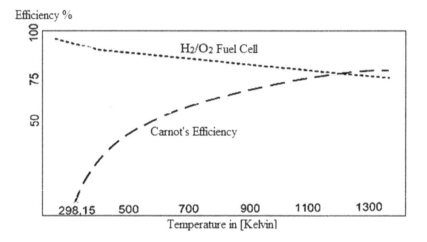

Figure 2.2. *Efficiency curve*

The curve in Figure 2.2 shows that the theoretical efficiency of the fuel cell is very high but, unlike thermal engines, it decreases when the temperature increases.

2.3. The flow rates of reactants and products

There is a relationship between the current delivered by the cell, the flow rate, the consumption of the reactants, and the production of water and heat. These parameters are called the "fuel cell flows" [BLU 07]. Figure 2.3 shows the input and outputs of a fuel cell.

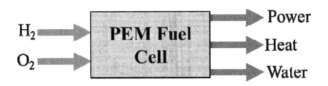

Figure 2.3. *Diagram of fuel cell flows*

2.3.1. *Oxygen flow rate*

According to the half-equation describing the reaction at the cathode:

$$O_2 + 4e^- + 4H^+ \rightarrow 2H_2O \qquad [2.47]$$

It is deduced that four electrons are transferred per mole of oxygen, the transferred charge will be:

$$Q = 4F \times \text{quantity of } O_2 \qquad [2.48]$$

In addition, the Faraday constant (F) is defined as the overall charge of one mole of elemental charge. By dividing it by time (J = dQ/dt), we obtain the molar flow rate (mol.s^{-1}) of oxygen as a function of the current. For an elementary cell of a fuel cell, we obtain:

$$\dot{n}_{O_2} = \frac{J}{4F} \qquad [2.49]$$

For a fuel cell with (n$_c$) cells:

$$\dot{n}_{O_2} = \frac{J}{4F} \cdot n_c \qquad [2.50]$$

It is sometimes preferable to use the raw power (P_{gross}) in the expression of oxygen flow rate. The power of the fuel cell is given by the expression:

$$P_{gross} = V \cdot J \cdot n_c \qquad [2.51]$$

and the expression of molar flow rate is:

$$\dot{n}_{O_2} = \frac{P_{gross}}{4 \cdot V \cdot F} \qquad [2.52]$$

For the mass flow rate (q_{O_2}) expressed in ($kg.s^{-1}$), we write:

$$q_{O_2} = \frac{M_{O_2} \cdot P_{gross}}{4 \cdot V \cdot F} = \frac{32 \times 10^{-3} \cdot P_{gross}}{4 \cdot V \cdot F} = 8.29 \times 10^{-8} \cdot \frac{P_{gross}}{V} \qquad [2.53]$$

where (M_{O_2}) is the molar mass of oxygen expressed in $kg.mol^{-1}$.

2.3.2. Hydrogen flow rate

The flow rate of hydrogen consumed is deduced in the same way as that of oxygen.

According to the half-equation describing the reaction at the anode:

$$2H_2 \rightarrow 4H^+ + 4e^- \qquad [2.54]$$

Note that two electrons are released per mole of hydrogen and the flow rate of hydrogen becomes:

$$\dot{n}_{H_2} = \frac{J}{2F} \cdot n_c \qquad [2.55]$$

and for the expression containing the gross power, we write:

$$\dot{n}_{H_2} = \frac{P_{gross}}{2 \cdot V \cdot F} \qquad [2.56]$$

For the mass flow rate (qH_2) expressed in ($kg.s^{-1}$), we write:

$$q_{H_2} = \frac{M_{H_2} \cdot P_{gross}}{2 \cdot V \cdot F} = \frac{2.02 \times 10^{-3} \cdot P_{gross}}{2 \cdot V \cdot F} = 1.05 \times 10^{-8} \cdot \frac{P_{gross}}{V} \qquad [2.57]$$

where (M_{H_2}) is the molar mass of oxygen expressed in $kg.mol^{-1}$.

2.3.3. Amount of water produced

According to the electrochemical half-equation, it can be seen that one mole of water is produced for two electrons. Therefore, water production in $mol.s^{-1}$ is:

$$\dot{n}_{H_2O} = \frac{P_{gross}}{2 \cdot V \cdot F} \qquad [2.58]$$

With the molar mass (MH_2O) being 18.02. 10^{-3} $kg.mol^{-1}$ and the production of water (in $kg.s^{-1} = 1$ $l.s^{-1}$ if we assume the water density is 1 $kg.l^{-1}$), the mass flow rate is therefore:

$$q_{H_2O} = \frac{M_{H_2O} \cdot P_{gross}}{2 \cdot V \cdot F} = \frac{18.02 \times 10^{-3} \cdot P_{gross}}{2 \cdot V \cdot F} = 9.34 \times 10^{-8} \cdot \frac{P_{gross}}{V} \qquad [2.59]$$

2.4. Electrochemistry of the fuel cell

The theoretical voltage (E_{th}) decreases once the battery is connected to an external circuit, and we must now calculate the actual voltage of the fuel cell. To do this, several phenomena must be considered such as the kinetics of the electrodes, the activation overpotential, and the currents and potentials that appear in the cell [OHA 09]. These phenomena describe the rate of the electrochemical reaction as well as the energy losses.

2.4.1. Electrode kinetics

When a metal electrode is in contact with an electrolyte, a characteristic potential difference is created between the internal potential of the electrode and the internal potential of the electrolyte.

This potential difference is referred to as the standard electrode potential (E^0) or the Galvani potential. The tables give the values of this measured potential difference with respect to a reference electrode (hydrogen electrode H_3O^+/H_2) whose potential is arbitrarily set to zero. If we have a system where the two electrodes are in contact with the same electrolyte, the voltage at the terminals of these electrodes, equal to the difference between the potentials of two electrodes, is denoted as (ΔE^0), under normal conditions of temperature and pressure (0°C and 1 atm or 273.15 K and 101 325 Pa). This voltage corresponds to an open circuit, therefore to a system at equilibrium. As soon as the electrons circulate, the system is out of equilibrium and the measured voltage (E) is different from the voltage (E^0). This difference is called the overpotential and it is indicated by the symbol (η) [BAR 05, BOU 07, HOR 09, SPI 07].

2.4.2. *Activation energy*

The surface of the electrodes is therefore the site of electrochemical reactions. These reactions result from the collision of the different molecules present. The kinetic energy of each molecule is the main source of energy to initiate a reaction. However, there is a minimum energy that must be reached in order for two molecules to react: the activation energy (energy threshold to cross). Part of the energy is lost because it is used to reach the activation level. The corresponding voltage loss is the activation overpotential mentioned (η_{act}). After collision, there is formation of an intermediate body (activated complex) which in turn decomposes to form the reaction products:

$$A + B \Leftrightarrow AB^* \Rightarrow C + D \qquad [2.60]$$

where AB^* is the activated complex.

This activation energy is calculated from the binding enthalpies. For the reaction between hydrogen and oxygen, for example, the necessary energies are those of dissociation of a molecule of hydrogen and a half-molecule of oxygen. From the tables of binding enthalpy values, we obtain:

$$H_2 \rightarrow 2H \text{ with } \Delta H = 436 \text{kJ} \cdot \text{mol}^{-1} \qquad [2.61]$$

and:

$$\frac{1}{2}O_2 \rightarrow O \text{ with } \Delta H = \frac{438}{2} kJ \cdot mol^{-1} \qquad [2.62]$$

that is, an activation energy of $436 + 219 = 685$ kJ.mol^{-1}.

2.4.3. Reaction rate

The use of a catalyst decreases the activation energy required for a reaction to take place by bringing into play intermediate reactions occurring on the surface of the catalyst.

Using Arrhenius' law, which gives the relation between reaction rate and temperature, it is possible to determine the effect of a variation in the activation energy (E_a) [BAR 05, HIR 98]:

$$k = A \cdot \exp\left(\frac{-E_a}{RT}\right) \qquad [2.63]$$

where (k) is the velocity constant, a function of temperature; and (A) is a constant, a function of the reaction. A small decrease in activationenergy causes a large increase in the reaction rate: the changing in activation energy from 100 to 90 kJ.mol^{-1} increases the reaction rate by approximately 60.

2.4.4. Exchange current

In an open circuit, even in the absence of current flow in a fuel cell, each electrode/electrolyte interface has a dynamic equilibrium and the charges (electrons) cross this interface in both directions. The resulting current density created by the flow of electrons is called the exchange current density (j_0), and it is expressed in amperes per square centimeter (A.cm^{-2}). This is the measure of the charge transfer rate at equilibrium. The higher it is, the easier the reaction is to initiate.

For electrochemical systems, (j_0) varies from a few nanoamperes per square centimeter to a few amperes per square centimeter [BOU 07]. The values of the exchange current density (j_0) for the H_2/O_2 system are summarized in Table 2.3.

	Catalyst	Environment	J_0 (A/cm²)
Hydrogen	Platinum	Acidic	10^{-3}
Electrolyte membrane	–	Basic	10^{-4}
Oxygen	Platinum	Acidic	10^{-9}

Table 2.3. *Current density values*

The exchange current depends on the concentration and also on temperature; the effective exchange current density (per unit of geometric surface area of the electrode). It is also a function of the charge (concentration) of the catalyst and the surface area of the electrode [GAS 03]. The following equation illustrates an expression of the exchange current density as a function of the reference value:

$$j = j_0^{ref} a_c L_c \left(\frac{P_r}{P_r^{ref}} \right)^\gamma \exp\left[-\frac{E_c}{RT}\left(1 - \frac{T}{T_{ref}} \right) \right]$$ [2.64]

where:

– j_0^{ref} is the reference exchange current density (at the reference temperature and pressure, typically 25°C and 101.25 kPa) per unit area of the catalyst (platinum "Pt" expressed in A.cm⁻² of Pt);

– a_c is the specific surface area of the catalyst (the theoretical limit of the platinum catalyst is 2,400 cm².mg⁻¹, between 600 and 1,000 cm².mg⁻¹ can be found; up to 30% of the platinum can be incorporated into the electrode;

– L_c is the catalyst charge (between 0.3 and 0.5 mg of Pt.cm⁻²), lower values are possible but they give rise to small voltages in the fuel cell;

– P_r is the partial pressure of the reactant (kPa);

– P_r^{ref} is the reference pressure (kPa);

– γ is the pressure coefficient (between 0.5 and 1.0);

– E_c is the activation energy, approximately 66 kj.mol⁻¹ (for the reduction of oxygen);

– R is the ideal gas constant (8.314 j. mol⁻¹ k⁻¹);

– T is the temperature (K);

– T_{ref} is the reference temperature (298.15 K).

2.4.5. *Current density*

Since the electrochemical reaction taking place in the cell is a surface reaction, the current density is therefore a characteristic variable of a fuel cell. It is also expressed in amperes per square centimeter. On the physical surface of the electrode, however, it is a function of many parameters (type of fuel cell, fuel flow rate, etc.) [BLU 07]. Out of equilibrium, the current density (j) is related to the activation overpotential (η_{act}) by the Butler–Volmer relation:

$$j = j_0 \left[\exp\left(\frac{\alpha n F}{RT} \eta_{act} \right) - \exp\left(-\frac{(1-\alpha) n F}{RT} \eta_{act} \right) \right] \qquad [2.65]$$

where (α) is the charge transfer coefficient which indicates the distribution of the activation overpotential between the anode and the cathode. It gives the proportion of the activation potential (η_{act}) required for the activation energy of anodic oxidation. The part of the activation energy required for the reduction at the cathode is therefore ($1-\alpha$). Its value is always between 0 and 1. For a symmetric reaction, $\alpha = 0.5$. For electrochemical reactions, the value of (α) is usually between 0.2 and 0.5 [BOU 07].

This equation, established by Butler and Volmer, describes, at the macroscopic level, the transfer kinetics of electrons in a heterogeneous medium. It is a simplified model of activation phenomena (assuming the charge concentration at the surface of the electrode is equal to that in the electrolyte, for example). It shows that the current density of a fuel cell increases exponentially with the activation overpotential (η_{act}) and that it is important to have a high exchange current density value (j_0), as it decreases the activation threshold by the use of a catalyst, by increasing the temperature or the concentration of reagents. For large overpotentials, $|\eta_{act}| \gg (RT/F)$, usually greater than 50 or 100 mV, the term corresponding to the opposite reaction is negligible (irreversible reaction) and we use the simplified equation, established empirically, known as the Tafel equation, which is a good approximation of the Butler–Volmer equation [BOU 07, LAR 03]. It allows the values (j_0) and (α) of the Tafel equation to be determined experimentally. This equation is as follows:

$$j = j_0 \left[\exp\left(\frac{\alpha n F}{RT} \eta_{act} \right) \right] \qquad [2.66]$$

For overpotential values below 10 mV, the Butler–Volmer equation can be simplified to a linear equation:

$$j \cong j_0 \left(\frac{nF}{RT} \eta_{act} \right)$$
[2.67]

2.5. Polarization phenomena

When the terminals of the fuel cell are connected to the charge traversed by a current of intensity (j), the voltage at the terminals of the fuel cell decreases with respect to the theoretical voltage due to polarization phenomena, of which there are three forms: activation, ohmic and concentration polarization.

2.5.1. *Activation polarization*

The activation polarization is greater at the cathode due to the kinetics of oxygen reduction, which is slower than the reduction of hydrogen at the anode.

This polarization is due to the activation energy of the reactions at the electrodes. The chemical kinetics at the electrodes is the result of complex reaction steps each with its own kinetics.

This activation polarization can be estimated using the empirical Tafel equation in its detailed form as follows [BLU 07, BOU 07, LAR 03, OHA 09]:

$$\eta_{act} = -\frac{RT}{\alpha nF} \ln j_0 + \frac{RT}{\alpha nF} \ln j$$
[2.68]

This equation gives the decrease in voltage at an electrode. In order to know the voltage decrease in a fuel cell, it is necessary to know the voltage decrease at the cathode and at the anode. Noting respectively (J_{0_a}) and (J_{0_C}), the exchange currents at the anode and at the cathode, and (A_a) and (A_c) the coefficients of the Tafel equation at the anode and at the cathode, the expression of the voltage decrease in the cell is:

$$\eta_{act} = A_a \ln \left(\frac{j}{j_{0\_a}} \right) + A_C \ln \left(\frac{j}{j_{0\_C}} \right)$$
[2.69]

The value (J_{0_c}) is much smaller, by a factor of 10^5, than the value of (J_{0_a}), it is therefore deemed that the overpotential at the anode is negligible; hence, the equation [2.69] can be written as follows:

$$\eta_{act} = A_{ac} \ln\left(\frac{j}{b}\right) \qquad [2.70]$$

with (A_{ac}) expressed in (V) and characterized by the following equation:

$$A_{ac} = A_a + A_c \qquad [2.71]$$

and (b) expressed in (mA/cm²) and characterized by the following relation:

$$b = j_{0\_a}^{\left(\frac{A_a}{A}\right)} + j_{0\_c}^{\left(\frac{A_c}{A}\right)} \qquad [2.72]$$

a being the slope of the Tafel curve:

$$A = \frac{RT}{\alpha nF} \qquad [2.73]$$

The curve in Figure 2.4 gives an example of the evolution of voltage during activation that a fuel cell can undergo.

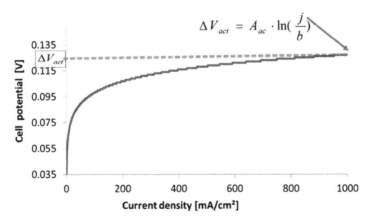

Figure 2.4. *Evolution of loss during activation*

2.5.2. *Ohmic polarization*

With increasing fuel flow rate, the current density increases but the internal resistance of the electrolyte causes a decrease in the voltage at the electrodes. There is a voltage decrease (η_{ohm}), also called ohmic losses, which is mainly a function of the electrical resistance of the membrane. These losses are due to conjugated phenomena, which include the electrical resistance of the electrode, the contact resistances and the resistance of the electrolyte to the passage of ions. The latter is characterized by the ionic conductivity (H^+) expressed by equation [3.84].

This resistance is actually dominant compared to the electrical resistance of the electrodes; it depends mainly on temperature. The voltage loss, caused by ohmic losses, is simply expressed by Ohm's law [SPI 07]:

$$\eta_{ohm} = R_m \cdot j \qquad [2.74]$$

where (R_m) is the Area Specific Resistance (ASR) of the membrane and it is expressed in ($\Omega.cm^2$). It depends on the ionic (protonic) conductivity of the membrane (H^+) and its thickness (e_m) expressed in (cm):

$$R_m = \frac{e_m}{\sigma_{H^+}} \qquad [2.75]$$

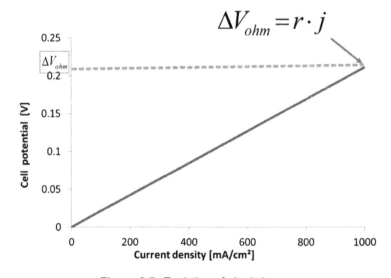

Figure 2.5. *Evolution of ohmic losses*

The straight line in Figure 2.5 shows the evolution of the ohmic voltage losses that a fuel cell can have.

The losses due to this ohmic polarization (also called resistance) can be minimized by reducing the resistance of the electrolyte (by decreasing the thickness, for example), or by increasing the ionic conductivity of the electrolyte (nature of the material, temperature increase, etc.).

2.5.3. *Concentration polarization*

Losses of concentration are due to the decrease in concentration of the reactants on the surface of the electrodes. They appear for high currents where the concentration of the reagents decreases greatly on the surface of the electrodes. In the case of air (which normally contains 0.21% oxygen), the fraction of oxygen at the surface of the electrodes will decrease significantly, depending on the current intensity, causing a drop in partial pressure of oxygen (the supply of oxygen is blocked by nitrogen) and therefore the voltage. In other words, it is the reaction rate that decreases. The same phenomenon occurs for the anode, which is supplied with hydrogen. The Nernst equation calculates the voltage loss if it goes from an initial concentration (arrival of reactants) to a final concentration. The extreme case being where the final concentration is equal to zero; in this case, we define a limiting current density (j_L) [BOU 07] and the concentration overpotential can be written in the following form:

$$\eta_{conc} = \frac{RT}{nF} \cdot \ln\left(\frac{j_L}{j_L - j}\right) \qquad [2.76]$$

The limiting current density (j_L) is a measure of the maximum flow rate of the gases; the current density (i) cannot be greater than (j_L). At this limiting current density, the concentration of the gases at the electrodes becomes almost zero and the voltage drops quite significantly.

Using the Butler–Volmer equation (reaction kinetics) leads to a similar result with a general equation in the form:

$$\eta_{conc} = c \cdot \ln\left(\frac{j_L}{j_L - j}\right) \qquad [2.77]$$

where (c) is a empirically obtained constant. We find in the literature [LAU 01], another formula, totally empirical but giving better results, expressing losses as concentration, shown by the following equation:

$$\eta_{conc} = m \cdot \exp^{(n \cdot j)} \tag{2.78}$$

where (m) expressed in (V) and (n) expressed in (cm²/mA) are empirical coefficients.

Another empirical equation suggested by Kim *et al.* [KIM 95] is as follows:

$$\eta_{conc} = c \cdot \exp^{\left(\frac{j}{d}\right)} \tag{2.79}$$

where (c) and (d) being empirical coefficients, with values $c = 3.10^5$ V and $d = 0.125$ A.cm² being suggested [LAR 03].

The curve in Figure 2.6 shows the evolution of voltage loss as a concentration that a fuel cell can have. The grouping of the different polarizations is presented in Figure 2.7.

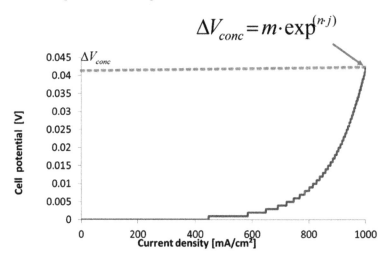

Figure 2.6. *Evolution of polarizations as concentration*

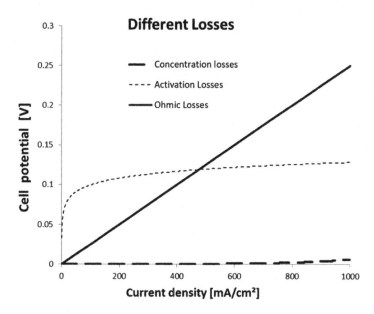

Figure 2.7. *The different voltage losses in a PEM fuel cell*

2.5.4. *Real cell voltage*

The combined effects of all the polarizations, also called irreversibilities or even dissipations, can be synthesized into a single equation giving the actual voltage of the fuel cell as a function of the current density (j) [HIR 98]:

$$V(j) = E_{theo} - \eta_{act} - \eta_{ohm} - \eta_{conc} \qquad [2.80]$$

2.5.5. *Polarization curve*

The main characteristic of a fuel cell can be summarized in the polarization curve. This curve is a graph representing the voltage at the terminals of the fuel cell as a function of the current intensity. A curve (I–V) is the most common method of characterizing or comparing one fuel cell to another. The polarization curve shows the voltage–current relationship as a function of the conditions of use such as humidity, temperature, electric charge, flow density of the fuel and the oxidant. Figure 2.8 shows a typical polarization curve of a PEM fuel cell, as well as all characteristic points.

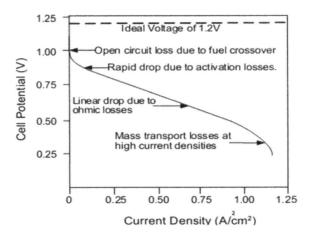

Figure 2.8. *Typical polarization curve of a PEMFC*

As shown in Figure 2.8, the polarization graph can be divided into three regions: activation, ohmic and concentration.

The operating range of a fuel cell is also characterized by three variables: voltage, efficiency and power [BOU 07].

In the stoichiometry, one mole of hydrogen (2 g) reacts with half a mole of oxygen (16 g). The theoretical electrical energy produced (G^0) is 237.34 kJ.mol^{-1}, or 66 kWh for a 2 g consumption of hydrogen. Below are some examples of orders of magnitude:

– for a light vehicle continuously requiring a power of 66 kW (about 90 hp), full-power operation would consume about 2,000 g of hydrogen per hour (distance traveled about 120 km), that is, a tank with a volume of about 60 L if the hydrogen is compressed at 350 bar;

– for a residential application (6.6 electric kWh required), the consumption will be about 200 g of hydrogen per hour;

– a power station of about 200 kWh of electricity would consume about 6 kg of hydrogen per hour;

– although the hydrogen mass seems to be low, the volume occupied for an optimal duration of operation is great, even if this hydrogen is significantly compressed, given that under normal conditions of temperature and pressure, 2 g of hydrogen occupies a volume of 22.414 L.

2.5.6. *Optimum operating range*

Polarization phenomena that lower the voltage when the current density increases force the fuel cell outside its limits (very low or very high current densities). In general, the optimal value is chosen in the linear zone. Given these phenomena and an external charge (motor, etc.), the actual voltage varies according to the required intensity, the type of fuel cell and the operating conditions (temperature, pressure, etc.); it is approximately 0.5 to 1 V. The fuel cell provides a variable voltage depending on the intensity [BOU 07].

2.6. Modeling of charge transfer

The performance of a steady-state PEMFC is generally expressed by the polarization curve. A steady-state model can be used to predict the polarization curve, and thus the fuel cell efficiency. The dynamic performance of the fuel cell can be modeled by an equivalent circuit that integrates a general solution of the steady-state model. Charge transfer models for PEMFCs can be divided into two broad categories, namely analytical and empirical. Analytical models are based on theory, whereas empirical models are based on experimentation. Unfortunately, PEMFC cells are complex systems and they are difficult to model completely in an analytical way. For this reason, analytical models of fuel cells have a theoretical basis while still maintaining certain empirical characteristics. Thirumalai and White [THI 97] supplemented, with a good sensitivity analysis, an old Nguyen and White [NGU 93] model, which aimed to find the best operating conditions for the reasonable operation of a PEMFC. They found that the following factors are of particular importance: gas mass flow, the operating temperature of PEMFC and the relative humidity of the reactive gases and more particularly that of hydrogen.

2.7. Overview of analytical models

Analytical models can be classified into two categories, namely simple and complex. The simple models characterize the operating voltage of the PEMFC while taking into account the theoretical maximum voltage as well as the major voltage losses.

Complex models have technical advantages and therefore require more information about the materials and constituents of the overall PEMFC.

2.7.1. *Simple analytical models*

These models attempt to represent the phenomena involved in a cell by simple and short equations describing the reversible voltage and voltage losses that may exist. Some initial studies, for example, that of Berger [BER 68], give a theoretical insight into the calculation methodology of the different voltage losses. Other groups of researchers aim to describe voltage losses while developing simple and accurate models. The most complete of these models, based on the study of a Ballard IV fuel cell, was developed by Amphlett *et al.* [AMP 95a, AMP 95b] who wanted to generalize the results of their work by a third approach in [MAN 00].

Other models, including that of Larminie and Dicks [LAR 03], describe the notion of internal current and include a simplified dynamic performance model of the fuel cell.

2.7.2. *Complex analytical models*

Complex analytical models must take into account many of the phenomena occurring in the fuel cell. In addition, the generation of these models requires a long period of expertise and observation of battery behavior under different conditions. The work of Murgia *et al.* [MUR 02], Bernardi and Verbrugge [BER 92], Nguyen and White [NGU 93] and Eikerling *et al.* [EIK 98] have been the subject of such a modeling category.

2.8. Empirical models

Due to the complexity and interdependence of the variables influencing the performance of the PEMFC, empirical equations could be used to predict the polarization curve. The advantage of this approach is that it is relatively simple to accurately predict the particular polarization curve. The disadvantage of this approach is that the polarization curve must be recalculated for any modification of operating conditions (such as humidity or temperature). An empirical equation (now a reference) was introduced in 1995 by Kim *et al.* [KIM 95] and gives a more accurate prediction of the polarization curve but neglects some electrochemical aspects. However, it is used in modeling the polarization curve for several types of PEMFC cells [ERI 01, PIS 02]. Other empirical models have been used, such as that of Amphlett *et al.* [AMP 95]; when the analytic constants are not known, the empirical equations for each parameter and for the overall ohmic loss will be used. Empirical modeling techniques are also used to develop an equivalent circuit to model the dynamic performance of the PEMFC. This technique uses experimental procedures such as Current interruption and Electrochemical Impedance Spectroscopy (EIS).

2.9. Current transport and charge conservation

In general, the current transport is described by the charge conservation equations [BAR 05, VAN 04] for the electric current:

$$\nabla \cdot \left(\sigma_s^{\mathrm{eff}} \cdot \nabla \varphi_s \right) = S_{\varphi s} \qquad [2.81]$$

For the ionic current:

$$\nabla \cdot \left(\sigma_m^{\mathrm{eff}} \cdot \nabla \varphi_m \right) = S_{\varphi s} \qquad [2.82]$$

where:

– σ_s^{eff} is the electrical conductivity in the solid phase, in [S.cm^{-1}];

– σ_m^{eff} is the electrical conductivity in the electrolyte (membrane), in [S.cm^{-1}];

– ϕ_s is the potential of the solid phase, in [V];

– ϕ_m is the potential of the membrane, in [V];

– S_ϕ is the source term representing the volumetric transfer current.

At the anode:

$$\begin{cases} S_{\phi s} = -j_a \\ S_{\phi m} = j_a \end{cases} \qquad [2.83]$$

At the cathode:

$$\begin{cases} S_{\phi s} = j_c \\ S_{\phi m} = -j_c \end{cases}$$ [2.84]

For the remaining areas of the fuel cell:

$$S_\phi = 0$$ [2.85]

For charge transport in diffusion layers, it is governed by Ohm's law:

$$\vec{j} = -\sigma_s^{eff} \cdot \vec{\nabla}\varphi_s$$ [2.86]

The charge retention is described in the electrodes by the following continuity relation:

$$\vec{\nabla}\vec{j} = 0$$ [2.87]

For charge transport in the membrane, the charge balance in the membrane is written as follows:

$$\vec{\nabla} \cdot \left(-\sigma_m \cdot \vec{\nabla}\varphi_m\right) = 0$$ [2.88]

2.10. Conclusion

In this chapter, we were able to highlight the different theoretical values of the reversible voltage and the voltage losses. Electrode overpotentials were also defined, estimated and then calculated from thermodynamic considerations. As for the actual voltage, it varies according to the required intensity, the type of fuel cell and the operating conditions (temperature, pressure, etc.); it is approximately 0.5 to 1.0 V.

The main characteristics of a fuel cell can be summed up as a polarization curve. This curve is a graph representing the voltage at the terminals of the fuel cell as a function of the current intensity. A curve (I–V) is the most common method of characterizing or comparing one fuel cell to another. The polarization curve shows the voltage–current relationship as a function of the conditions of use such as humidity, temperature, electric charge, the flow density of the fuel and the oxidant. Polarization phenomena, which lower the voltage when the current density increases, force the fuel cell outside its limits (very low or very high current densities).

Finally, several analytical, empirical or semi-empirical models are presented and compared to give an insight into the work already done on mathematical models aimed at studying the performances (static or dynamic) of the PEM fuel cell.

Mass Transfer Phenomena

3.1. Introduction

A fuel cell should be supplied continuously with fuel (hydrogen) and an oxidant (oxygen) to ensure the production of electricity. On the contrary, the products of the reaction in the heart of the cell should also be released continually for maximum efficiency, as well as to improve its lifetime. Studying the mass transfer of the different species involved in the fuel cell is essential. In fact, poor mass flow management can lead to significant failures in the performance of the battery.

3.2. Flow of matter

Concentrations of reactants and products adjacent to the catalyst layers have a strong influence on determining the performance of the cell. Decreases in concentration can be minimized by optimizing the upstream transport of these species, from flow channels to electrodes and diffusion layers (GDL). Table 3.1 summarizes the role of mass transfer in each cell component of a PEMFC [SPI 07].

Two main mass transfer phenomena are encountered in the PEMFC, diffusion in the diffusion layers (GDLs) and electrodes, and convection in the flow structures (flow channels); this can be related to the difference in dimensions. Indeed, the length of the flow structures is in millimeters, or even centimeters, and the structure is a well-defined set of channels. For the diffusion layers and electrodes, the porous structures are in micrometers or even nanometers, and the tortuous structures trap the molecules of the reactive gases and thus the convective force coming from the flow structures is slowed, favoring transport by diffusion from the GDLs. Figure 3.1 represents a summary diagram of the various mass transfer phenomena involved at the heart of a PEMFC.

Components	Implications of mass transfer	Existing limitations
Flow channels (H_2/O_2)	Ensure homogeneous distribution of reagents across the electrode surface, while minimizing pressure loss and maximizing the ability to evacuate water.	Depletion of reagents. Contamination of the catalyst sites with impurities (N₂N, for example).
GDLs	Allows easy access of gases to catalytic layers (electrodes) and improves thermal conductivity.	Liquid (non-evacuated) water blocks the pores of the catalyst (the electrode).
The electrodes	The electrochemical reaction takes place at the catalyst interface; it consumes the reagents (H_2/O_2) and produces water.	Slow reaction if the catalyst charge is low at the electrodes.
The membrane	Good water management for good ion transport (H^+ protons).	Drying of the membrane (at high temperatures), which results in the loss of its proton conductive capacity.

Table 3.1. *The role of mass transfer in a PEM fuel cell*

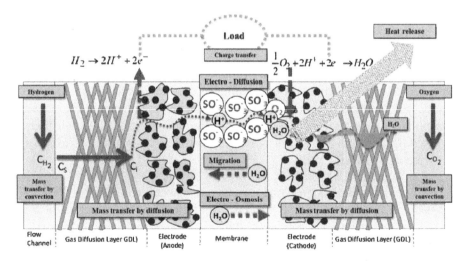

Figure 3.1. *Descriptive diagram of the mass transfer phenomenon*

The convective forces that dominate the mass transfer in the flow channels are mainly imposed by the flow of fuel (hydrogen), while the flow conditions of the oxidant can be controlled by the user. High reagent flow rates can provide good distribution at the catalytic sites of the electrodes, but this can also cause problems in the overall fuel cell and can even damage the polymer membrane. The diffusive forces acting on the surface of the electrodes are supported by convective forces upstream (from the flow channels). Indeed, the velocity of the reagents decreases along its path to the diffusion layers where diffusion begins. The consumption of gaseous reactants and the production of water at the cathode are a function of the flow of electrons passing through the external circuit, and therefore of the intensity, through the electrochemical reactions. In other words, the material flow densities are directly related to the current density delivered by the fuel cell [RAM 05]. The diagram in Figure 3.2 shows a general overview of material flows. In what follows, we will consider the case of steady state and unidirectional mass transfer (Ox) for a PEMFC.

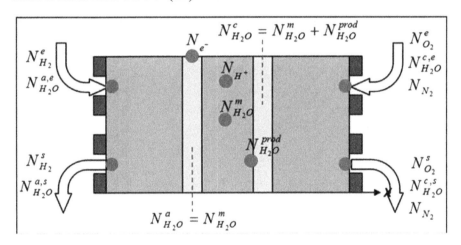

Figure 3.2. *The different material flows in a PEM fuel cell*

The current density (j) expressed in [A.cm^{-2}] is related to the molar electron flow density, we then write:

$$N_{e^-} = \frac{j}{F}$$

[3.1]

We recall the equations of the electrochemical reaction at the electrodes:

– anode:

$$H_2 \longrightarrow 2H^+ + 2e^-$$ [3.2]

– cathode:

$$\frac{1}{2}O_2 + 2H^+ + 2e^- \longrightarrow H_2O$$ [3.3]

By respecting the stoichiometric coefficients, the quantities of material produced and consumed at the electrodes per unit area and time and as a function of the current density, let us recall molar flow density equations, but this time for a single cell and according to the current density.

For the flows consumed:

– in $x = L_d$, we write:

$$- N_{H_2} = \frac{j}{2F}$$ [3.4]

– in $x = L_d + L_m$, we write:

$$- N_{O_2} = \frac{j}{4F}$$ [3.5]

For the flows produced: in $x = L_d + L_m$, we write:

$$- N_{H_2}^{Prod} = \frac{j}{2F}$$ [3.6]

where (L_d) and (L_m) correspond to the respective thicknesses of the diffusion layers and the membrane. In addition, the steady-state hypothesis assumes the absence of species accumulation; it is given by the relationship:

$$\forall j, \forall x \qquad \frac{dN_j}{dx} = 0$$ [3.7]

By convention, flows are positive when they move from the anode to the cathode; the location of the wells (and sources) of matter imposes the densities of flows of matter that are written as follows:

– at the anode:

$$N_{H_2} = \frac{j}{2F} \qquad\qquad [3.8]$$

– in the membrane:

$$N_{H^+} = \frac{j}{F} \qquad\qquad [3.9]$$

– at the cathode:

$$N_{O_2} = -\frac{j}{4F}; \quad N_{N_2} = 0 \qquad\qquad [3.10]$$

It should be noted that the flow of nitrogen is considered zero because it is not involved in the reactions, and it is also assumed that the membrane is impervious to this gas.

In addition, the continuous flow of water at the interfaces (membrane/diffusion layers) must be taken into account:

– at the anode (in $x = L_d$), we have:

$$N^m_{H_2O} = N^a_{H_2O} \qquad\qquad [3.11]$$

– at the cathode (in $x = L_d + L_m$), we have:

$$N^c_{H_2O} = N^m_{H_2O} + N^{Prod}_{H_2O} \qquad\qquad [3.12]$$

The water flow density ($N^m_{H_2O}$) is mainly imposed by the transfer mechanisms in the membrane, whereas the other densities of material flow are known as a function of the current density [RAM 05].

3.3. Mass transfer by convection

The reactive gases in the flow channels have the concentration (C_0). They will be transported, by convection, to the diffusion layers to be well distributed at the surface of the electrodes. The concentration then becomes (C_s). Figure 3.3 shows the mass transfer phenomena along the anode side of the fuel cell [SPI 07].

Convective mass transfer for this fuel cell is written as [LI 06]:

$$m = A_{elec}h_m(C_0 - C_S)$$ [3.13]

where:

– A_{elec} is the electrode surface;

– h_m is the mass transfer coefficient.

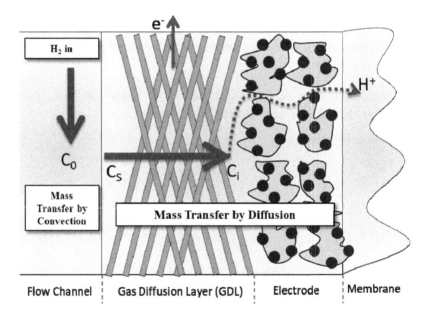

Figure 3.3. *Mass transfer at the anode*

The value of (h_m) depends on channel geometry and physical properties of species (i) and (j), it can be determined from the Sherwood number [SPI 07]:

$$h_m = Sh \frac{D_{i,j}}{D_h}$$

[3.14]

where:

– D_h is the hydraulic diameter;

– $D_{i,j}$ is the binary diffusion coefficient of species (i) and (j);

– Sh is the Sherwood number.

The latter depends on the geometry of the channels, it can have the value of (Sh = 5.39) for a uniform mass flux on the surface ($\dot{m}$ = constant), and the value of (Sh = 4.86) for a uniform concentration on the surface of the electrodes (C_S = Constant).

The dependence of the binary diffusion coefficient ($D_{i,j}$) on temperature can be expressed as [BAR 05]:

$$D_{i,j}(T) = D_{i,j}(T_{ref}) \cdot \left(\frac{T}{T_{ref}} \right)^{3/2}$$

[3.15]

where:

– T_{ref} is the temperature with which the binary diffusion coefficient is given;

– T is the fuel temperature used in the fuel cell.

The production of water at the cathode is shown in Figure 3.4 [SPI 07]. The water produced at the cathode will be eliminated by convective airflow.

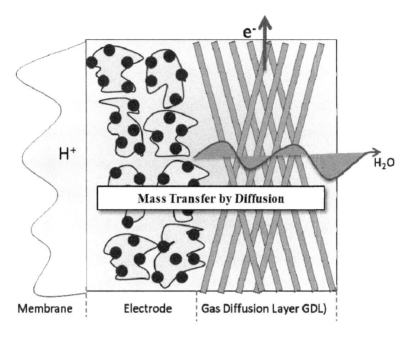

Figure 3.4. *Production of water at the cathode*

3.4. Mass transfer in porous diffusers

3.4.1. *Conservation of mass*

The generalized mass conservation equation, valid for the description of mass transfer phenomena inside a fuel cell (flows, diffusion, phase change and electrochemical reactions) [LAL 12], is written as:

$$\frac{\partial}{\partial t}\left(\rho \cdot \varepsilon_s\right) + \vec{\nabla} \cdot \left(\rho \cdot \vec{u}\right) = 0 \qquad [3.16]$$

where:

– ε_s is the porosity of the medium (ratio between the pore volume and the total volume);

– ρ is the density of the gas mixture [kg.m^{-3}];

_ $\vec{u}$ is the superficial rate [m.s^{-1}].

Darcy's law, which describes the convection of a gas mixture in a porous medium (diffusion layers), is based on the homogenization of porous and fluid media in a place where the detailed geometric description of the pore structure is not necessary [GAL 05]. In this case, the volume of the medium is considered relatively large in relation to the pore size; therefore, the speed of the gas mixture is determined by the pressure gradient, the viscosity and the structure of the porous medium [NGU 10]:

$$\vec{u} = -\frac{k_P}{\mu}\vec{\nabla}p \qquad\qquad [3.17]$$

where:

– μ is the dynamic viscosity of the gas [kg.m^{-1}.s^{-1}];

– K_p is the permeability [m^2];

– P is the pressure [Pa].

3.4.2. The conservation of species

The species conservation equation representing conservation of mass for the gas phase of a single species is written as [BAR 05]:

$$\frac{\partial(\rho \cdot \varepsilon \cdot w_i)}{\partial t} + \vec{\nabla}\left(\rho \cdot \varepsilon \cdot w_i \cdot \vec{u}\right) = \vec{\nabla}\left(\rho \cdot D_i^{eff} \cdot \vec{\nabla}w_i\right) + S_{s,i} \qquad [3.18]$$

where:

– w_i is the mass fraction of the species (i) in its gaseous phase (H_2, O_2, H_2O and N_2);

– $S_{s,i}$ is the source term (or well) of the species (i);

– D_i^{eff} is the diffusion coefficient of the species (i) in a porous environment in [m^2.s^{-1}];

– $\dfrac{\partial(\rho \cdot \varepsilon \cdot w_i)}{\partial t}$ is the species accumulation term;

– $\vec{\nabla}\left(\rho \cdot \varepsilon \cdot w_i \cdot \vec{u}\right)$ is the advection term;

– $\vec{\nabla}\left(\rho \cdot D_i^{\text{eff}} \cdot \vec{\nabla} w_i\right)$ is the Fick's term for the diffusion of species in a porous medium.

The source term ($S_{s,i}$) is cancelled out everywhere in the cell with the exception of the catalytic layer where the species are consumed or generated during the electrochemical reaction. In this case, the source term is given by the following equations [BAR 05]:

– for H_2:

$$S_{s,H_2} = -i_a \cdot \frac{M_{H_2}}{2F} \qquad [3.19]$$

– for O_2:

$$S_{s,O_2} = -i_c \cdot \frac{M_{O_2}}{4F} \qquad [3.20]$$

– for H_2O (vapor):

$$S_{s,H_2O(g)} = \delta . A_{fg} \cdot \left(w_{sat} - w_{H_2O(g)}\right) \qquad [3.21]$$

– for H_2O (liquid):

$$S_{s,H_2O(l)} = +i_c \cdot \frac{M_{H_2O}}{2F} - \delta . A_{fg} \cdot \left(w_{sat} - w_{H_2O(g)}\right) \qquad [3.22]$$

where:

– δ is the evaporation coefficient, in $[\text{kg.m}^{-2}.\text{s}^{-1}]$;

– A_{fg} is the phase change area per volume unit, in $[\text{m}^{-1}]$;

– w_{sat} is the maximum mass fraction of water vapor in the dry gas (at saturation), in $\left[\text{kg}_{H_2O} \cdot \text{kg}_{gas}^{-1}\right]$;

– $w_{H_2O(g)}$ is the mass fraction of the water vapor in the dry gas, in $\left[\text{kg}_{H_2O} \cdot \text{kg}_{gas}^{-1}\right]$.

In the term source of water, it is assumed that the produced water is in the liquid state, and then it becomes vapor if the medium (airflow channel or O_2) is not saturated [BAR 05].

A bibliographic synthesis from [BER 91, COS 01, DJI 02, FUL 93, GUR 98, SPR 91] allows us to see how these authors have described the diffusion of gases (binary H_2 and H_2O at the anode and ternary of O_2, N_2 and H_2O at the cathode) in the diffusion layers (anode and cathode) using the Stefan–Maxwell model which is a generalization of Fick's law for two or more constituents [SPR 91], and that takes into account the collisions between the gas molecules [RAM 05]. This model makes it possible to express the variations in molar concentrations (C_i) of the constituents of the gas mixture as a function of their molar flux densities (N_i) with ($i = 1...n$), it is still defined as the gradient of the molar fraction for each species [AMU 03, BIR 02, CUR 99]:

$$\begin{cases} \vec{\nabla} x_i = \sum_{j=1, j\neq i}^{N} \left(\frac{1}{c \cdot D_{ij}^{\text{eff}}} \left(x_i \cdot \vec{N}_j - x_j \vec{N}_i \right) \right) \\ \sum_{i=1}^{N} x_i = 1 \end{cases}$$

[3.23]

where:

$-N$ is the number of species ($N = 2$ for the anode and $N = 3$ for the cathode);

$- \vec{N}_i$ is the molar flow density of the gas (i) in [mol.m^{-2}.s^{-1}];

$- x_i$ is the molar fraction of the gas (i), with ($x_i = c_i/C$);

$-c$ is the total concentration of the mixture in [mol.m^{-3}] with ($c = \sum_{i=1}^{n} c_i$);

$- D_{ij}^{\text{eff}}$ is the binary diffusion coefficient of the gaseous mixture in a porous medium of species (i) in species (j), in [m^2.s^{-1}].

From the ideal gas law, the total pressure is related to the concentration of the mixture:

$$p = c \cdot R \cdot T \tag{3.24}$$

Due to the porous structure of the electrodes, the binary diffusivity calculation needs to be corrected by the porosity (ε_s) of the medium [BER 92]:

$$D_{ij}^{eff} = D_{ij} \cdot \varepsilon_s^\tau \tag{3.25}$$

The effective diffusion coefficient (D_{ij}^{eff}), which is generally weaker than the intrinsic diffusion coefficient in an exclusively fluid phase (D_{ij}), is not only a function of the porosity (ε_s) of the medium, but also a function of the tortuosity given by the Bruggeman correlation ($\tau=1.5$) [AMA 03]. The porous and tortuous structure is responsible for lengthening the particle path and causes a decrease in the diffusion coefficient [BER 92]. It should also be noted that the diffusion coefficients of gases (i) and (j) are linked by a reciprocal relationship, here known as the Onsager reciprocal relation [MAZ 03]:

$$D_{ij}^{eff} = D_{ji}^{eff} \tag{3.26}$$

The transport of species in the diffusion layers (GDL) includes the Stefan–Maxwell binary diffusion and convection by Curtis and Bird [CUR 99]. It is expressed as a mass conservation equation of each of the components of the gas mixture [NGU 10]:

$$\frac{\partial}{\partial t}\left(\rho \cdot \varepsilon \cdot w_i\right) + \vec{\nabla} \cdot \left(-\rho \cdot w_i \cdot \sum_{j=1, j \neq i}^{N} D_{ij} \cdot \left[\vec{\nabla} x_j + \left(x_j - w_j\right) \cdot \frac{\vec{\nabla} p}{p}\right] + \rho \cdot w_i \cdot \vec{u}\right) = 0 \tag{3.27}$$

with i = H_2 or H_2O for the anode, and i = N_2, O_2 or H_2O for the cathode. The density of the gas mixture is determined by:

$$\rho = \frac{p / (RT)}{\displaystyle\sum_{i=1}^{N} \frac{w_i}{M_i}} \tag{3.28}$$

The mass fraction (w_i) of the gas (i) is calculated from the molar fraction (x_i) and the molar mass of the gases as follows [NGU 10]:

$$\begin{cases} w_i = \dfrac{x_i \cdot M_i}{\displaystyle\sum_{j=1}^{N} x_j \cdot M_j} \\[3mm] \displaystyle\sum_{i=1}^{N} w_i = 1 \end{cases} \qquad [3.29]$$

For a good description of mass transfer in the binary mixture at the anode and ternary at the cathode, the system of coupled differential equations must be solved.

The transfer is considered to be unidirectional according to (Ox), and that the system is isobaric [RAM 05]; therefore:

– at the anode:

- the gaseous mixture is binary; the species involved are hydrogen and water vapor, so the equation system to solve is written as:

$$\begin{cases} x_{H_2} + x^a_{H_2O} = 1 \\[3mm] \dfrac{dx_{H_2}}{dx} = \dfrac{1}{c \cdot D^{eff}_{H_2,H_2O}} \cdot \left[x_{H_2} \cdot \left(N^a_{H_2O} + N_{H_2} \right) - N_{H_2} \right] \end{cases} \qquad [3.30]$$

- at the diffuser inlet, the relative humidities (RH) in the gaseous mixture introduced into the bipolar plate channels are expressed as follows:

$$\begin{cases} x^{ae}_{H_2O} = HR_a \cdot \dfrac{p_{sat}}{p} \\[3mm] x^e_{H_2} = 1 - HR_a \cdot \dfrac{p_{sat}}{p} \end{cases} \qquad [3.31]$$

The exponent " e " refers to the input supply conditions of the fuel cell;

– at the cathode:

- the gas mixture is ternary, the species involved are oxygen, nitrogen and water vapor; the equation system to solve is written as:

$$
\begin{cases}
x_{O_2} + x_{N_2} + x^c_{H_2O} = 1 \\[2mm]
\dfrac{dx_{N_2}}{dx} = -\dfrac{1}{c}\left(\dfrac{N_{O_2}}{D^{eff}_{O_2,N_2}} + \dfrac{N^c_{H_2O}}{D^{eff}_{H_2O,N_2}}\right)\cdot x_{N_2} \\[4mm]
\dfrac{dx_{O_2}}{dx} = -\dfrac{1}{c}\left[\dfrac{N_{O_2} + N^c_{H_2O}}{D^{eff}_{O_2,H_2O}}\cdot x_{O_2} + \left(\dfrac{1}{D^{eff}_{O_2,H_2O}} - \dfrac{1}{D^{eff}_{O_2,N_2}}\right)\cdot N_{O_2}\cdot x_{N_2} - \dfrac{N_{O_2}}{D^{eff}_{O_2,H_2O}}\right]
\end{cases}
\qquad [3.32]
$$

- at the diffuser inlet, the relative humidities (RH) are expressed as follows:

$$
\begin{cases}
x^{ce}_{H_2O} = HR_c \cdot \dfrac{p_{sat}}{p} \\[3mm]
x^{e}_{O_2} = 0,21\left(1 - HR_c \cdot \dfrac{p_{sat}}{p}\right) \\[3mm]
x^{e}_{N_2} = 0,79\left(1 - HR_c \cdot \dfrac{p_{sat}}{p}\right)
\end{cases}
\qquad [3.33]
$$

3.4.3. *Some parametric laws*

3.4.3.1. *Saturation pressure*

Different formulas have been proposed for determining the saturation pressure of water (in [Pa]). Wöhr presented, in his work, the following empirical formula [MON 06]:

$$
P_{sat}(T) = 10^{\left(A - \frac{B}{C+T-273}\right)} \cdot 100
\qquad [3.34]
$$

In this relation, the temperature is in Kelvin, and the coefficients have the following values: A = 8.073; B = 1 656.39; C = 226.86 in K. Chun-Ying Hsu *et al.* [WÖH 00] used another formula where the decimal logarithm of

(P_{sat}) is obtained from a temperature polynomial of order 3. In this case, the unit of (P_{sat}) is [atm]:

$$\log\left(P_{sat}\left(T\right)\right) = -2.1794 + 0.02953 \times \left(T - 273\right) - 9.1837 \times 10^{-5} \times \left(T - 273\right)^2$$
$$+ 1.4454 \times 10^{-7} \times \left(T - 273\right)^3$$

[3.35]

3.4.3.2. *Binary diffusion coefficient of the gas mixture*

The effective diffusion coefficient in a porous medium of the binary mixture of gases (i) and (j) is expressed in [$m^2.s^{-1}$], and it is determined from the following empirical equation [HSU 09]:

$$D_{ij} = D_{ij}^0\left(T_0, P_0\right) \cdot \frac{P_0}{P} \cdot \left(\frac{T}{T_0}\right)^{1.5}$$

[3.36]

The reference pressure (P_0) is 1 atm, and the reference temperature (T_0) is chosen according to the gas pair considered. The different values of (D_{ij}^0) are given in Table 3.2.

Binary gas	T_0 [K]	Unit	Value
D_{H_2,H_2O}^0	307.1		$9.15 . 10^{-5}$
D_{O_2,H_2O}^0	308.1		$2.82 . 10^{-5}$
D_{O_2,N_2}^0	293.2	$m^2.s^{-1}$	$2.2 . 10^{-5}$
D_{N_2,H_2O}^0	307.5		$2.56 . 10^{-5}$

Table 3.2. *The values of the binary diffusion coefficients*

Another relationship presented in the literature is expressed as follows [BIR 02]:

$$D_{ij}^{eff} = \frac{a}{p}\left(\frac{T}{\sqrt{T_{C,i} \cdot T_{C,j}}}\right)^b \cdot \left(p_{C,i} \cdot p_{C,j}\right)^{1/3} \cdot \left(T_{C,i} \cdot T_{C,j}\right)^{5/12} \cdot \left(\frac{1}{M_i} + \frac{1}{M_j}\right)^{1/2} \cdot \varepsilon^{1.5} \quad [3.37]$$

where:

- T_C is the critical temperature of (i) or (j);

- p_c is the critical pressure of (i) or (j);

- M is the molecular mass of (i) or (j);

- a is 0.0002745 for diatomic gases (H_2, O_2 and N_2);

- a is 0.000364 for liquid H_2O;

- b is 1.832 for diatomic gases (H_2, O_2 and N_2);

- b is 2.334 for H_2O vapor.

3.4.3.3. Sorption isotherm

The water, liquid in the membrane and in as a vapor in the diffusers, undergoes a phase change at the membrane/diffuser interface (at the electrodes); this phenomenon is known as sorption.

Indeed, the water vapor molecules will bind to the structure of the membrane, where they will condense. This phenomenon is called adsorption (or simply sorption); and the reverse phenomenon is called desorption [SER 07]. The sorption mechanism is responsible for a release of heat. The porous structure of the polymer site allows the adsorption of existing water into the gas diffusion layers in gaseous form [GOT 08, ZAW 93]. Assuming there is thermodynamic equilibrium between the water vapor in the support layers and the liquid water in the polymer, the sorption curve of Hinatsu *et al.* [HIN 94] enables the water content of (λ_a) and (λ_c) to be calculated, according to the activity of the water vapor at 30 and 80°C [NGU 10]:

$$\lambda_{30°C} = 0.043 + 17.81 \times a - 39.85 \times a^2 + 36 \times a^3 \qquad [3.38]$$

$$\lambda_{80°C} = 0.3 + 10.8 \times a - 16 \times a^2 + 14.1 \times a^3 \qquad [3.39]$$

The activity of the water (in the vapor state) in the diffusion layers is calculated from the water concentration and the saturation pressure at the relevant cell temperature at the membrane/diffuser interface [NGU 10]:

$$a = c_{H_2O} \cdot \frac{RT}{P_{sat}(T)} \qquad [3.40]$$

For a temperature between 30 °C and 80 °C, the water content is estimated by interpolating these two curves in a linear fashion (see Figure 3.5). The value of (P_{sat}) is a function of the temperature according to the following correlation [RAM 05]:

$$\frac{P_{sat}}{P} = \exp\left(13.669 - \frac{5096.23}{T_{moy}}\right) \qquad [3.41]$$

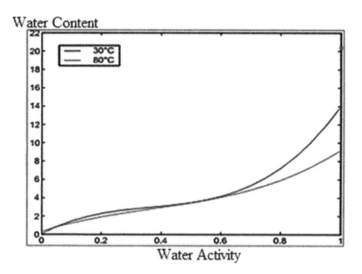

Figure 3.5. *Sorption isotherms*

These isotherms reflect the equilibrium between the water vapor that bathes the membrane and the liquid water it contains.

3.5. Mass transfer in the catalyst layers (electrodes)

The catalyst layers (CLs) are sites in which electrochemical reactions occur. Due to the sintering process, part of the catalyst is in the membrane. However, two models can be distinguished to describe the mass transfer in the catalyst layers: the Butler–Volmer low current model and the high current model, also known as the agglomerate model.

3.5.1. *Low current model (Butler–Volmer)*

Since these layers are very thin compared to the other elements of the fuel cell (channels, GDL), they are considered as interfaces, or boundary conditions, in the mathematical models. In order to obtain the distribution of the local current density on the catalyst surface, the Butler–Volmer kinetic equation is used in static and dynamic models [BAR 01, BER 02]. At very low currents (at least several µA), this equation provides a coherent simulation fuel cell voltage (unlike the agglomerate model), that is, the fuel cell voltage remains below the equilibrium voltage [COL 08]. Thus, the current densities are calculated as follows [BER 02]:

– at the anode:

$$j_a = j_a^0 \left(\frac{c_{H_2}}{c_{H_2}^{ref}} \right)^{0,5} \cdot \left[\exp\left(\frac{\alpha_a \cdot F}{RT} \cdot \eta_a \right) - \exp\left(-\frac{\alpha_C \cdot F}{RT} \cdot \eta_a \right) \right] \qquad [3.42]$$

– at the cathode:

$$j_c = j_c^0 \left(\frac{c_{O_2}}{c_{O_2}^{ref}} \right) \cdot \left[\exp\left(\frac{\alpha_a \cdot F}{RT} \cdot \eta_c \right) - \exp\left(-\frac{\alpha_c \cdot F}{RT} \cdot \eta_c \right) \right] \qquad [3.43]$$

where:

– α_a, α_c is the charge transfer coefficients;

– c_{H_2}, c_{O_2} is the hydrogen and oxygen concentrations at the membrane/electrode interfaces.

They are calculated using Henry's law:

$$c_{H_2} = \frac{P_a \cdot x_{H_2}}{K_{H_2}} \qquad [3.44]$$

and:

$$c_{O_2} = \frac{P_c \cdot x_{O_2}}{K_{O_2}} \qquad\qquad [3.45]$$

K_{H_2}, K_{O_2} are the Henry constants of hydrogen and oxygen. In equations [3.42] and [3.43], the activation overpotentials $(_{a,c})$ in [V] are determined from the local potentials of the membrane electrodes:

$$\eta_{a,c} = \phi_{a,c} - \phi_m - E_{a,c}^{eq} \qquad\qquad [3.46]$$

where $E_{a,c}^{eq}$ is the equilibrium potential of the anode and the cathode [WAN 06]. The equilibrium potential of the anode is zero, and that of the cathode depends on the fuel cell temperature (in Kelvin) according to the expression:

$$E_c^{eq} = 1.23 - 0.9 \times 10^{-3} \times (T - 298) \qquad\qquad [3.47]$$

3.5.2. Agglomerate model with strong current

To integrate the transport of the reactants in the activation layers into the simulation, Gloaguen et al. [GLO 98] noted that the agglomerate model was more accurate than the macro-homogeneous model. As a result, this agglomerate model describes two processes of transporting the reactants to the catalyst sites. The first process presents the diffusion of gases through the secondary pores between the agglomerates. The second integrates the dissolution of the reactants in the electrolyte before reaching the reaction layers. Applying this type of model, Siegel et al. [SIE 03] showed that the performance of the fuel cell depended considerably on the catalyst structure. Ion current densities $(j_{a,c})$ in [A.m^{-2}] are related to the average current densities $(i_{agg}^{a,c})$ on the surface of the agglomerate particles [MAH 06, SHI 06]:

$$j_{a,c} = L_{act} \cdot (1 - \varepsilon_{CL}) \cdot j_{a,c}^{agg} \qquad\qquad [3.48]$$

where:

– L_{act} is the thickness of the reaction layer in [m];

– ε_{CL} is the porosity of the reaction layer.

Current density densities ($j_{a,c}^{agg}$) in [A.m^{-3}] are determined by applying the diffusion equation and the Butler–Volmer electrokinetic equation for agglomerates [BAR 01].

At the anode:

$$j_a^{agg} = -6 \cdot n_a^{e^-} \cdot F \cdot \left(\frac{D_{H_2}^{agg}}{R_{agg}^2} \right) \cdot \left(c_{H_2}^{agg} - c_{H_2}^{ref} \cdot \exp\left(\frac{2F}{RT} \cdot \eta_a \right) \right)$$

$$\cdot \left(1 - R_{agg} \cdot \sqrt{\frac{j_a^0 \cdot S_a}{2 \cdot F \cdot c_{H_2}^{ref} \cdot D_{H_2}^{agg}}} \cdot \coth\left(R_{agg} \cdot \sqrt{\frac{j_a^0 \cdot S_a}{2 \cdot F \cdot c_{H_2}^{ref} \cdot D_{H_2}^{agg}}} \right) \right)$$

[3.49]

where:

– $D_{H_2}^{agg}$ is the diffusion coefficient of hydrogen in agglomerates, in [m^2.s^{-1}];

– R_{agg} is the radius of the agglomerate, in [m].

The concentration of hydrogen in the anode reaction layer is that given by equation [3.44]:

$$c_{H_2}^{agg} = c_{H_2}$$

[3.50]

S_a represents the specific surface area per unit volume in each agglomerate particle. The number of charges for the anode is ($n_a^e = 1$):

$$j_c^{agg} = -6 \cdot n_c^{e-} \cdot F \cdot \left(\frac{D_{O_2}^{agg}}{R_{agg}^2} \right) \cdot c_{O_2}^{agg} \cdot$$

$$\left(1 - R_{agg} \cdot \sqrt{\frac{j_c^0 \cdot S_a}{2 \cdot F \cdot c_{O_2}^{ref} \cdot D_{O_2}^{agg}} \cdot \exp\left(-\frac{0.5 \times F}{R \cdot T} \cdot \eta_c \right)} \cdot \right.$$
$$\left. \coth\left(R_{agg} \cdot \sqrt{\frac{j_c^0 \cdot S_a}{2 \cdot F \cdot c_{O_2}^{ref} \cdot D_{O_2}^{agg}} \cdot \exp\left(-\frac{0.5 \times F}{R \cdot T} \cdot \eta_c \right)} \right) \right)$$
[3.51]

here, with ($n_c^e = 2$). The oxygen concentration in the anode reaction layer is that of equation [3.45]:

$$c_{O_2}^{agg} = c_{O_2}$$
[3.52]

Exchange current densities ($j_{a,c}^0$) in [A.m^{-2}] express the reaction rates of the anode and cathode, respectively. We could also introduce the source terms ($j_{a,c}^{agg}$) in the evaluation of the divergence of current densities at the electrodes [BIR 02, ZIE 05]. It should also be noted that at very low currents (several µA or less), the current densities generated in the activation layers can be calculated using the relations [3.49] and [3.51], which give atypical overvoltages (the resulting battery voltage is greater than the equilibrium voltage) [BOU 07].

3.6. Mass transfer in the membrane

The membrane must transfer the protons from the anode to the cathode. As this membrane is hydrated, the transfer of water must be taken into account as well.

The transfer of water occurs primarily by the transfer mechanisms in the membrane. The resistive effects to the transfer of the species in the fuel cell are much greater in the electrolyte (diffusion of a liquid) than in the diffusers (gaseous phases, ensuring a better diffusion).

To ensure this, we can first calculate the diffusion resistance in each medium, not taking into account the convective effects.

This comparison has already been done by Ramousse [RAM 05], and it is described as follows: for the diffusion of water in the diffusers, we chose the weakest effective diffusion coefficients used for our modeling $(D^{eff}_{H_2O,O_2})$.

The diffusive transfer resistances in the diffusers and in the membrane therefore have for values (for $L_d = 230$ µm and $L_m = 175$ µm):

$$\begin{cases} D^{diff}_{H_2O} \approx 3\times10^{-5}; in \left[m^2 \cdot s^{-1} \right] \\ R^{diff} = \dfrac{L_d}{D^{diff}_{H_2O}} \approx 7.7 ; in \left[s \cdot m^{-1} \right] \\ D^{memb}_{H_2O} \approx 3\times10^{-9}; in \left[m^2 \cdot s^{-1} \right] \\ R^{memb} = \dfrac{L_m}{D^{memb}_{H_2O}} \approx 6\times10^{4}; in \left[s \cdot m^{-1} \right] \end{cases} \qquad [3.53]$$

It is therefore clear that the transfer into the membrane will be more limiting than into the diffusers [RAM 05]. The characterization of the mass transfer phenomenon in the electrolyte deserves our attention, especially since these electrical performances, and therefore those of the cell, are strongly dependent on it.

3.6.1. *Schröeder's paradox*

Values, experimentally recorded, show a strong discontinuity between the cases of a membrane immersed in saturated steam and saturated liquid water. When the membrane is immersed in liquid water, its water content increases considerably. For example, at 80 °C, the water content is ($\lambda = 20.4$) in liquid water and only ($\lambda = 9.2$) in saturated vapor. Choi and Datta [CHO 03] explained this difference by the surface tensions at the interface between the vapor and liquid phases, which reduce the amount of vapor water adsorbed. This apparent paradox may be due to the absence of measurements corresponding to a fluid phase partially saturated with liquid (with a = 1).

The discontinuity of the amount of water in the membrane in the vicinity of the saturation results in a discontinuity of the water flow in the membrane, which is not compensated by the continuous production.

The transport of water in the membrane is described by the phenomenological model given by the references [BER 92, BER 91, NGU 93, SPR 91, SPR 93]. It is the combination of two movements: diffusion, created by the gradients of water concentration in the membrane, and the other is electro-osmotic, generated by the clusters of water molecules carried by protons $(H_2O)_n H^+$ when they cross the membrane from the anode towards the cathode.

The number of water molecules carried by a proton (see [3.76]), denoted (ξ), is known as the water transport coefficient by the electro-osmotic effect.

The molar flow density of the water transported in the membrane is defined in $[mol.m^{-2}.s^{-1}]$ and it is written as:

$$\vec{N}_{H_2O}^m = \xi \cdot \frac{-\sigma_{H^+} \cdot \vec{\nabla}\phi_m}{F} - D_{H_2O}^m \cdot \vec{\nabla}c_{H_2O}$$
[3.54]

where:

– ϕ_m is the membrane potential [V];

– σ_{H^+} is the ionic conductivity of the membrane $[S.m^{-1}]$;

– $D_{H_2O}^m$ is the diffusion coefficient of water in the membrane in $[m^2.s^{-1}]$.

These last two parameters strongly depend on the water content; the conservation equation of the amount of water is written as:

$$\frac{\partial c_{H_2O}}{\partial t} + \vec{\nabla} \cdot \vec{N}_{H_2O}^m = 0$$
[3.55]

3.6.2. Microscopic scale

In the literature, we find modeling of the water transport in the membrane according to three scales: microscopic, mesoscopic and macroscopic. The different models found in the literature are detailed. At the microscopic scale, where the studied systems are smaller than 100 atoms, techniques based on statistical mechanics and molecular dynamics are used to describe the transfer mechanisms. These techniques consider all the atoms of a molecular system and only a pore fragment is studied.

Kreuer *et al.* [KRE 04] synthesized simulation tools and proton transfer mechanisms in membranes.

3.6.2.1. *Statistical mechanics*

Explaining the behavior of macroscopic systems from their microscopic characteristics is the subject of statistical mechanics. It thus makes it possible to study the Brownian motion of the particles (individually monitoring their position in a fluid) and it is based on a main assumption that the equations of classical mechanics are still valid (Langevin principle).

It is then possible to study the interactions between the different molecules present in the membrane pores.

Paul and Paddison [PAU 04] used this simulation technique to evaluate the relative permittivity of water (ε_r) in a cylindrical pore filled with water and ions by formulating a hypothesis of local thermodynamic equilibrium. They deduced radial variations for (ε_r) in the pore. It should be noted that a study based on statistical mechanics requires the introduction of probability densities, thus making the overall study more complex [COL 08].

3.6.2.2. *Molecular dynamics*

This is another very small-scale modeling technique that simulates the evolution of a particle system over time (hence, "dynamics").

The trajectory of a molecule is calculated by applying the laws of Newtonian classical mechanics.

These simulations are carried out for a few nanoseconds, during which the time is discretized. The system is composed of a hundred molecules; each molecule is considered as a dynamic entity for which the position of atoms evolves over time. At each time step, Newton's second law makes it possible to know the speed and position of each atom by knowing the forces acting on it.

In the case of Nafion membranes, there are four types of forces: the interactions between the atoms, the forces at the interfaces, the permanent stresses (e.g. the temperature) and the driving forces (gravity, gradients of pressure and concentration) [DIN 98].

The way the atomic interaction forces (or the potentials from which these forces derive) are determined characterizes the simulation type. If the potentials are calculated precisely from the principles of quantum mechanics, we refer to molecular dynamics (*ab initio*).

On the contrary, if the forces derive from a predetermined potential, fixed empirically or determined by the calculation of an electronic structure, we then refer to molecular dynamics (*a priori*). This method has been used mainly to study the transport of protons in Nafion.

According to [BRA 07, CUI 07, DIN 98, EIK 02, ELL 07, LI 01, PAD 01, PAD 03, WES 06], with this method, the motions of the atoms can be seen as vibrations around an energy minimum. Their determination, for each molecule, makes it possible to obtain fundamental information on the local structure of the membrane, the proton diffusion, the aggregation of ionic groups (through the formation of hydrogen bonds), or even the dissociation of protons in the sulfonate groups.

Din and Michaelides [DIN 98], for example, conducted a study in surface-filled cylindrical pores, in which flowed an electrolytic solution (water + protons).

The pore radius is around one nanometer, and the surface charge density (κ) in the pore is constant and equal to (-0.1) or (-0.2) [$C.m^{-2}$].

Overall, the authors find that a high concentration of non-hydrated protons (low volume) is adsorbed near the charged boundary, while a second layer of hydrated (and therefore larger) protons is observed within a few angstroms of the surface. The higher the surface charge density (κ), the more the number of ions attracted to the boundary increases.

Finally, when the water content in the pore is high, the proportion of hydrated protons is greater.

The information obtained by molecular dynamics simulations are numerous and rich. In general, the influence of parameters such as the temperature, the water content, the ion exchange capacity or the physical and chemical characteristics of the main and side chains has been studied in the literature [COM 02, DOK 06, SEE 05, SPO 02].

However, equations of motion can only be solved by means of complex numerical methods [COL 08].

3.6.3. *Mesoscopic scale*

Mesoscopic modeling is the application of macroscopic transport laws to the microscopic structure of the membrane. This kind of modeling uses the Poisson–Boltzmann theory and it has been the subject of great attention.

The Poisson–Boltzmann theory is used by Gross and Osterle [GRO 68] to describe the interactions between the electrostatic field and the ion concentration in the pore.

They have thus developed a model of coupled transfers in a charged cylindrical pore with a uniform charge surface density.

Knowing the charge density (ρ_e), the Poisson equation calculates the electrostatic potential (ϕ) in the pore [BOU 07]:

$$\vec{\nabla}(\varepsilon_r \cdot \vec{\nabla}\phi) = -\frac{\rho_e}{\varepsilon_0} \qquad [3.56]$$

with:

$$\rho_e = F\sum_i z_i \cdot c_i^m \qquad [3.57]$$

where c_i^m is the concentration of the mobile ionic species (i) with charge (z_i) in the pore, it is given by the statistic distribution law of Boltzmann:

$$c_i^m(r) = c_i^b \cdot \exp\left(-\frac{z_i \cdot F\phi(r)}{RT}\right) \qquad [3.58]$$

with (c_i^b), the concentration of the ionic species in an external solution or in a bulk is the point of the solution where the electroneutrality relation is verified: the effects of the positive and negative charges cancel each other out; the electrostatic potential of the double layer is zero. This equation takes into account the attraction/repulsion forces between the protons and the charged surface. In order to complete the description of the phenomena at the pore scale, here are some authors: Koter [KOT 00, KOT 02], Cwirko and Carbonell [CWI 92a, CWI 92b] and Pintauro *et al.*

The authors of the research [BON 94, GUZ 90, TAN 97, YAN 00, YAN 04] use a modified Boltzmann equation to determine the ion concentration. This equation, developed by Gur *et al.* [GUR 78] and used by Verbrugge and Hill [VER 88], takes into account the effects of dielectric saturation in the vicinity of a charged surface [BOU 07]:

$$c_i^m(r) = c_i^b \cdot \exp\left(-\frac{z_i \cdot F\phi(r)}{RT} - \frac{\Delta^m G_i(r) - \Delta^b G_i}{RT} \right) \qquad [3.59]$$

where G_i is the variation in Gibbs free energy of the ionic species considered to be between a studied medium and the vacuum. The second term of the exponential takes into account the hydration forces acting on the protons. These forces exist because the interactions between the ions and the solvent are not uniform in a charged pore. Theoretical studies to determine the hydration forces were conducted by Bontha and Pintauro [BON 94] and by Eikerling and Kornyshev [EIK 01]. Another modification of the classical Poisson–Boltzmann theory is obtained by taking into account the confinement of water in regions with sizes of only a few nanometers (typically, a Nafion membrane pore). High electrostatic fields can significantly reduce the relative permittivity of water (ε_r) compared to that of the bulk (ε_b) [CWI 92a, CWI 92b].

This is the case of the charged pore wall edges. This effect has been considered by many authors [BON 94, CWI 92a, CWI 92b, GUZ 90, KOT 02, TAN 97, YAN 00, YAN 04] using the Booth equation [BOO 51]:

$$\varepsilon_r = n^2 + 3\frac{\left(\varepsilon^0 - n^2\right)}{\beta_B \cdot \nabla\phi} \cdot \left[\frac{1}{\tanh\left(\beta_B \cdot \nabla\phi\right)} - \frac{1}{\beta_B \cdot \nabla\phi} \right] \qquad [3.60]$$

with:

$$\beta_B = 5 \cdot \frac{\mu_d}{2 \cdot k_b \cdot T} \cdot \left(n^2 + 2\right) \qquad [3.61]$$

where (n) is the refractive index of the solvent, (μ_d) is the dipole moment of the solvent and (k_b) is the Boltzmann constant. Examination of this equation shows that (ε_r) decreases as the electrostatic field increases. The molecules of the solvent are polarized, orientated towards the charged pore wall and their mobility is reduced. The work of Paul and Paddison [PAU 04] and Paddison [PAD 01] confirmed this evolution using statistical mechanics. However, no experimental information is available in the literature concerning the arrangement of water molecules under these conditions. The classical or modified Poisson–Boltzmann theory has therefore often been used to describe the spatial distribution of protons in the pore. The solvent is considered as an incompressible continuous fluid whose displacement is governed by the Navier–Stokes equation. According to Choi *et al.* [CHO 05, CHO 06], the ion transport mechanisms at the atomic scale are well described in terms of diffusion by the Nernst–Planck equation. Solving these equations makes it possible to elucidate the transport in the ion exchange membrane. Assuming a constant relative water permittivity (ε_r) in the pore, Gross and Osterle have linked the ion concentration and electrostatic potential profiles in the pore to the transport coefficients of the theory of irreversible process thermodynamics. Verbrugge and Hill [BOO 51] were the first to show the relevance of the predictions of their model by comparing them with their experimental data. Cwirko and Carbonell [CWI 92a, CWI 92b] have confirmed the good model/experiment concordance based on the results of Narebska [NAR 84]. More recently, Yang and Pintauro [YAN 04] developed a model with a variable pore size. They show that the hydration forces greatly influence the spatial distribution of the ions: the hydrated ions are excluded from the areas close to the charged wall because of their real size. In all cases, the numerical results of the models are compatible with the experimental data recorded by the various authors.

Nevertheless, a model based on these equations (in particular, the modified Poisson–Boltzmann equation) needs to be solved numerically and it is difficult to introduce it into an overall fuel cell model [BOU 17].

3.6.4. *Macroscopic scale*

3.6.4.1. *Modeling transport in a porous medium*

This model is based on the description of Gierke [GIE 82]; it is also called hydraulic model and it comes from the theoretical work of Pintauro and Verbrugge [PIN 89].

The authors of this model propose a quantitative description of the equilibrium of a membrane immersed in an acidic solution. Indeed, the membrane is considered to be made up of channels with impermeable negatively charged walls in which liquid water and protons circulate. Liquid water and protons form a charged solution that is subject to electrical potential and hydraulic pressure gradients.

The principle of this modeling is to link the electrical potential distribution to ion concentration profiles in the pores of the membrane [CWI 92b].

The distribution of the electric potential and the transport of ions and water are respectively described by the Poisson–Boltzmann, Nernst–Planck and Navier–Stokes equations. The transport coefficients, hydraulic permeability, electrokinetic permeability and ionic conductivity are usually derived from experimental measurements or can be evaluated from data such as swell, ion exchange capacity, pore size and physical properties of the solution that surrounds the membrane [KOT 02]. The validity range of this type of description is set at a pore size of around 10 times the size of a water molecule, that is, approximately 3 nm.

This approach, based on a microscopic description of pore transport phenomena (double layer type), allows the modeling of macroscopic transport mechanisms by the application of a scaling method, such as taking an average, for example. This type of model has been used for many years for transport in clays and membranes, a synthesis of which is presented in work by Lemaire [LEM 04], and it is recently applied to the case of a Nafion membrane, particularly by Bernardi and Verbrugge [BER 92], and Singh and Djilali [SIN 99].

For the transport of water, as formulated by Bernardi and Verbrugge, the hydraulic model assumes the coexistence of a solid phase (membrane) and a liquid phase (water + protons). The solid phase is supposed to be inert; the displacement of the charged solution in the membrane is the result of the action of an electric field and a pressure gradient. The proton water solution in the membrane has a low flow velocity. To describe the transport of water in the membrane, we can then use the Stokes equations where the gravitational forces are negligible compared to the electric forces.

The average rate of the fluid phase $\langle \vec{u} \rangle$ is given as a function of the pressure and potential gradients, in the macroscopic multidimensional case by the Schlögl equation. In addition to the Darcy term for the pressure gradient, there is a term that translates the effects of an electrical potential gradient on the charged solution (water + protons). This is known as electro-osmosis:

$$\langle \vec{u} \rangle = -\frac{K_p}{\mu} \vec{\nabla} P_b - \frac{K_E}{\mu} \cdot C_f \cdot F \cdot \vec{\nabla}\varphi \qquad [3.62]$$

where:

– $\langle \vec{u} \rangle$ is the average flow rate, in (m.s^{-1});

– μ is the dynamic viscosity of water, in (kg .m^{-1}.s^{-1});

– φ is the macroscopic electric potential, in (V);

– C_f is the fixed charge concentration, in (mol.m^{-3});

– P_b is the pressure of the liquid water in the pore, in (Pa);

– K_p and K_E are respectively the hydraulic and electrokinetic permeabilities, in Darcy (D) with $(1D = 0.97 .10^{-12} \, m^2)$.

For the transport of ions in the membrane, the protons migrate by convection (overall movement of the solution) and by diffusion under the effect of an electrochemical potential gradient ($\tilde{\mu}_{H^+}$). At the microscopic scale, the Nernst–Planck relationship is used to determine the proton flow (N_{H^+}):

$$N_{H^+} = c_{H^+} \cdot \vec{u} - D_{H^+,H_2O} \cdot \vec{\nabla}\tilde{\mu}_{H^+} \qquad [3.63]$$

where:

– D_{H^+,H_2O} is the effective diffusion coefficient (protons in water) of the pores of the membrane, in (m².s^{-1});

– c_{H^+} is the proton concentration, in (mol.m^{-3});

– $\vec{u}$ is the microscopic velocity of water in the pores of the membrane, in $(m.s^{-1})$;

– $\mu_{H^+}^-$ is the electrochemical potential of the protons.

This chemical potential is written for a charged particle in an electric field:

$$\mu_{H^+}^- = F\varphi + RT \ln c_{H^+} \qquad [3.64]$$

If the concentration (c_{H^+}) is assumed to be uniform in the membrane, only an electrical potential gradient induces diffusion displacement of the charged particles. We are talking about migration; equation [3.63] can be rewritten in the following form:

$$N_{H^+} = c_{H^+} u - D_{H_2O,H^+}^{eff} \cdot \frac{F \cdot c_{H^+}}{RT} \nabla \varphi \qquad [3.65]$$

Averaged over the membrane, Bernardi and Verbrugge [BER 92] wrote the previous equation as follows:

$$\langle N_{H^+} \rangle = \langle c_{H^+} \rangle u - D_{H_2O,H^+}^{eff} \cdot \frac{F \cdot \langle c_{H^+} \rangle}{RT} \nabla \varphi \qquad [3.66]$$

In their model, Bernardi and Verbrugge assume that the solvent velocity is uniform over a straight section of a pore. Consequently, $\langle c_{H^+} u \rangle = \langle c_{H^+} \rangle \cdot u$. In addition, electroneutrality is respected in the pore and it becomes $\langle c_{H^+} \rangle = C_f$. It is then possible to rise to steady state; the average current density across the membrane $\langle j \rangle = F \langle N_{H^+} \rangle$ given by the authors in [BER 91] is:

$$\langle j \rangle = -\frac{K_E}{\mu} \cdot \vec{\nabla} P_b - \delta_{H^+} \cdot \vec{\nabla} \varphi \qquad [3.67]$$

where:

- K_E is the electrokinetic permeability, in Darcy (D);

- δ_{H^+} is the ionic conductivity of the membrane, in $(S. \ m^{-1})$.

By associating the equations [3.62] and [3.66], we write:

$$j = -F \langle c_{H^+} \rangle \cdot \frac{K_p}{\mu} \cdot \nabla P_b - \left[F^2 \langle c_{H^+} \rangle \frac{K_E}{\mu} \cdot C_f + D^{eff}_{H_2O,H^+} \cdot F^2 \cdot \frac{\langle c_{H^+} \rangle}{RT} \right] \cdot \nabla \varphi \quad [3.68]$$

Ion conductivity is defined by the ratio $(i/\nabla \phi)$:

$$\delta_{H^+} = \underbrace{F^2 \langle c_{H^+} \rangle \frac{K_p}{\mu} \cdot C_f}_{\text{Convective Term}} + \underbrace{D^{eff}_{H_2O,H^+} \cdot F^2 \cdot \frac{\langle c_{H^+} \rangle}{RT}}_{\text{Diffusive Term}} \quad [3.69]$$

In the case of Nafion membranes, the fluid flowing in the pores contains charged particles (protons). Their convective displacement also contributes to the current density. The ionic conductivity is then the sum of a diffusive term and a convective term; the convective term exists for this model, because it is assumed that the protons are mainly present in the membrane, with the OH⁻ being negligible. In the case of a symmetrical electrolytic solution (with as many co-ions as of counter-ions), this convective term vanishes and the ionic conductivity depends solely on the diffusive term [BOU 07]:

$$\delta_{H^+} = D^{eff}_{H_2O,H^+} \cdot F^2 \cdot \frac{C_f}{RT} \quad [3.70]$$

Bernardi and Verbrugge only take into account the diffusive term of conductivity, despite the convective displacement of protons contributing to the current density. This hypothesis would only be valid in the case where the velocity of the fluid is zero: $\langle u \rangle = 0$.

To use the hydraulic model in a global fuel cell model, it is necessary to know the boundary conditions. Bernardi and Verbrugge use this transport model for a membrane bathed in liquid water. This implies a continuous liquid pressure at the interfaces [BOU 07].

In this case, the membranes are probably very hydrated. If the membranes are less hydrated, Weber and Newman [WEB 04a] suggested to describe the transport of the species in the membrane using the so-called diffusive model, which is presented below.

For Onsager's reciprocity relation, the two macroscopic transport laws previously formulated for the transport of charge (Nernst–Planck) and matter (Schlögl) in the membrane verify Onsager's reciprocity relations [RAM 05]:

$$\begin{pmatrix} \langle u \rangle \\ \langle j \rangle \end{pmatrix} = - \begin{pmatrix} \dfrac{K_P}{\mu} & \dfrac{K_E}{\mu} \\ \dfrac{K_E}{\mu} & \delta_{H^+} \end{pmatrix} \cdot \begin{pmatrix} \nabla P_b \\ \nabla \varphi \end{pmatrix}$$

[3.71]

On the contrary, the equations used by Bernardi and Verbrugge [BER 91] do not respect these relations. Indeed, the authors started from writing the Nernst–Planck relationship in the form:

$$\langle N_{H^+} \rangle = \langle c_{H^+} \rangle \langle u \rangle - D_{H^+,H_2O} \cdot \nabla \langle \mu_{H^+}^- \rangle$$

[3.72]

Whereas scaling methods require the expression of the convective transport term as follows $\langle c_{H^+} \cdot u \rangle$, with $\langle c_{H^+} \cdot u \rangle \neq \langle c_{H^+} \rangle \cdot \langle u \rangle$.

3.6.4.2. Phenomenological model

Unlike the hydraulic model, the diffusive model assumes the membrane is formed of a single homogeneous phase. Several models are based on this hypothesis including Springer [SPR 91], Fuller and Newman [FUL 93] and Okada [OKA 98]. This theory was transformed into an equation for Nafion membranes by Pintauro and Bennion [PIN 84].

The model of Fuller and Newman [FUL 93] is based on the theory of concentrated solutions, and the Springer model [SPR 91] is based on the theory of dilute solutions which considers only the interactions between dissolved species (water and protons) and the solvent (membrane) [NEW 91].

This formalism has also been taken up more recently by Janssen [JAN 01] as well as Weber and Newman [WEB 04b]. However, Meyers [MEY 98] remarks that it is not possible in this case to characterize the transport in a membrane with a pressure gradient at its boundaries. In general, the membrane can thus be assimilated to an electrolytic solution, where the transport of water is governed by two contributions:

– a diffusive Fick's flow generated by the water concentration gradients in the membrane. This flow can be indifferently directed from the anode to the cathode or vice versa depending on the humidification conditions of the membrane;

– an electro-osmotic flow describes the procession of water molecules carried by each proton when crossing the membrane. This flow, proportional to the proton flow (i/F), is always directed from the anode to the cathode.

According to Okada, the transport by electro-osmosis is the result of two simultaneous effects [OKA 98]:

– an electrostatic effect ensuring the solvation of the protons $(H_2O)_nH^+$ in hydronium (H_3O^+);

– a volume effect due to the size of the solvated molecules that will push the water molecules.

In the theory of dilute solutions, the transport of species results from migration, diffusion and convection phenomena. The flow of the species (i) is given by [BOU 07]:

$$N_i = -z_i \cdot \frac{D_i}{RT} \cdot F \cdot c_i \cdot \nabla \varphi - D_i \nabla c_i + c_i v \qquad [3.73]$$

where:

– z_i is the valency (electroneutrality) of species (i);

– v is the velocity of the solvent.

In this case, the solvent corresponds to the single phase of the membrane and therefore ($v = 0$); convective transport is canceled out (unlike the hydraulic model).

In the absence of a proton concentration gradient in the membrane, proton transport by diffusion also vanishes, and equation [3.73] is then limited to Ohm's law:

$$j = -\delta_{H^+} \cdot \nabla\phi \qquad [3.74]$$

with (δ_{H^+}) always being the ionic conductivity of the membrane:

$$\delta_{H^+} = \frac{z_i \cdot D_i \cdot F^2 \cdot c_i}{RT} \qquad [3.75]$$

Unlike the hydraulic model, the displacement of the charged solution does not contribute to the current density. On the contrary, the flow of protons induces a movement of water in the same direction: the electro-osmotic flow. This flow results from interactions between protons and water molecules. In the literature, an electro-osmotic coefficient (ξ) is defined as the number of water molecules accompanying each proton during its transport [OKA 98]:

$$\xi = \frac{n_{H_2O}}{n_{H^+}} \qquad [3.76]$$

The electro-osmotic flow ($N_{H_2O}^{Osmosis}$) is directly proportional to the proton flow in the membrane and it is expressed by:

$$N_{H_2O}^{Osmosis} = \xi \cdot \frac{j}{F} \qquad [3.77]$$

Water is an electrically neutral species ($Z_{H_2O} = 0$). According to equation [3.73], the transport of water also occurs by diffusion and the diffusive flow ($N_{H_2O}^{Diff}$) is given by:

$$N_{H_2O}^{Diff} = -D_{H_2O}^m \cdot \nabla c_{H_2O} \qquad [3.78]$$

where:

– $D_{H_2O}^m$ is the diffusion coefficient of the water in the membrane;

– c_{H_2O} is the water concentration in the membrane, it is related to the water content of the membrane (λ) by relation [3.81].

Springer *et al.* [SPR 91] ultimately assume that the flow of water (N_{H_2O}) in the membrane is the sum of electro-osmotic and diffusive flows:

$$N_{H_2O} = N_{H_2O}^{Osmosis} + N_{H_2O}^{diff}$$
$$(N_{H_2O})$$

[3.79]

The Springer model has been widely used in the PEMFC literature. In the case of Nafion membranes, transport parameters (σ_{H^+}, ξ and $D_{H_2O}^m$) are usually expressed as a function of water content (λ) and temperature (T).

A literature review shows that there is a multitude of expressions for transport coefficients, which have been presented in the papers [COS 01, DUT 01, FUL 92, FUL 93, ISE 99, KUL 03, MEI 04, NEU 99, REN 97, REN 01, SPR 91, YEO 77].

For example, for Nafion membranes, ionic conductivity (σ_{H^+}) is approximately 0.1 S.cm^{-1}, the electro-osmotic coefficient (ξ) is between 1 and 5, and the water diffusion coefficient in the membrane ($D_{H_2O}^m$) is approximately 10^{-9} m^2.s^{-1}. Nevertheless, there is a strong disparity between the different values of the transport coefficients (σ_{H^+}, ξ and $D_{H_2O}^m$) and this can lead to heterogeneous results.

The comparison of the results of the model with the experimental results makes it possible to judge its relevance. This task was carried out by Springer and his team [SPR 93], who were able to validate this second approach for several operating conditions of a fuel cell.

This modeling is relatively simple to implement; however, it involves a water transport coefficient by electro-osmosis and a diffusion coefficient generally evaluated experimentally. These coefficients, difficult to evaluate, largely determine the transport of water in the membrane.

3.6.5. *Parametric laws*

3.6.5.1. *Water content*

The diffusion coefficient of water in the membrane ($D_{H_2O}^m$) strongly depends on the local water content, which is defined as the ratio between the number of water molecules and the number of sulfonic sites (SO_3) available in the polymer. The following descriptions involve the water content of the membrane (λ). This variable represents the number of water molecules per active site (SO_3):

$$\lambda = \frac{n_{H_2O}}{n_{SO_3^-}} \qquad\qquad [3.80]$$

In addition, electroneutrality imposes an equal number of positive charges (protons) and negative charges (active sites). The ratio of water molecules to active sites then reflects the potential number of water molecules in the vicinity of each proton. The water content can also be defined according to the equivalent mass (EW) in (g mol$^1_{SO_3}$), the density of the dry membrane ($_{dry}$) expressed in (g.m^{-3}), and the water concentration in the membrane (c_{H_2O}):

$$\lambda = \frac{EW}{\rho_{dry}} \cdot c_{H_2O} \qquad\qquad [3.81]$$

In practice, the value of (λ) varies between 2 and 22.

In this relation, we neglect the variation of volume related to the variation of concentration in water. The variable (λ) is commonly used in the literature because it makes it possible to determine the hydration state of the membrane with respect to a controlled parameter, the number of active sites. The water content can also be expressed using the coefficient of water activity as follows:

$$\lambda = \begin{cases} 0.043 + 17.81a - 39.85\,a^2 + 36\,a^3 & (0 < a \le 1) \\ 14 + 1.4(a-1) & (1 < a \le 3) \\ 16.8 & (a > 3) \end{cases} \quad [3.82]$$

with:

$$a = \frac{P_{H_2O}}{P_{sat}} = \frac{P \cdot x_{H_2O}}{P_{sat}} \quad [3.83]$$

3.6.5.2. The ionic conductivity of the membrane

Work by Siegel [SIE 03] and Ge [GE 03] propose, for the ionic conductivity (of protons H^+) as a function of the water content, and in the case where ($\lambda > 1$), the following empirical determination:

$$\sigma_{H^+} = (0.5139 \cdot \lambda - 0.326) \cdot \exp\left[1268 \cdot \left(\frac{1}{303} - \frac{1}{T} \right) \right] \quad [3.84]$$

On the contrary, in an article by Neubrand *et al.* [NEU 99], we find the following formula, valid irrespective of the value of (λ):

$$\sigma_{H^+} = (0.2658 \cdot \lambda + 0.0298 \cdot \lambda^2 + 0.0014 \cdot \lambda^3) \cdot$$
$$\exp\left[(2640 \cdot \exp(-0.6 \times \lambda) + 1183) \cdot \left(\frac{1}{303} - \frac{1}{T} \right) \right] \quad [3.85]$$

3.6.5.3. The diffusion coefficient of water in the membrane

The water diffusion coefficient in the membrane depends on the type of membrane. For a Nafion membrane, A.C. West and T.F. Fuller [WES 96] gave, in their study, the following formula:

$$D_{H_2O}^m = 3.5 \times 10^{-7} \cdot \exp\left(-\frac{2436}{T} \right) \cdot \lambda \quad [3.86]$$

Motupally [MOT 00] presented another formula for determining this water transport coefficient as a function of (λ):

$$D_{H_2O}^m = \begin{cases} 3.1\times10^{-7}\cdot\lambda\cdot\left[\exp(0.28\times\lambda)-1\right]\cdot\exp\left(\dfrac{-2346}{T}\right) & \text{if } \lambda\leq3 \\[3mm] 4.17\times10^{-8}\cdot\lambda\cdot\left[1+161\cdot\exp(-\lambda)\right]\cdot\exp\left(\dfrac{-2346}{T}\right) & \text{if } \lambda>3 \end{cases} \qquad [3.87]$$

Finally, again for Nafion membranes, Neubrand *et al.* [NEU 99] have adopted the following empirical formula:

$$D_{H_2O}^m = 10^{-10.775+0.3436\cdot\lambda-0.189\cdot\lambda^2+0.0004\cdot\lambda^3}\cdot$$
$$\exp\left[\left(2640\cdot(-0.6\times\lambda)+1517\right)\cdot\left(\frac{1}{353}-\frac{1}{T}\right)\right] \qquad [3.88]$$

A few years ago, a study presented a relationship derived from experience for estimating the coefficient ($D_{H_2O}^m$) in a Gore-Select membrane [YE 07]:

$$D_{H_2O}^m = 0.5\times10^{-10}\cdot\exp\left[2416\cdot\left(\frac{1}{303}-\frac{1}{T}\right)\right]\cdot$$
$$\left(2.563-0.33\cdot\lambda+0.0264\cdot\lambda^2-0.000671\cdot\lambda^3\right) \qquad [3.89]$$

3.6.5.4. *The electro-osmotic coefficient*

The coefficient (ξ), involved in equation [3.54], has been experimentally measured by Springer *et al.* [SPR 91] for a Nafion membrane:

$$\xi = \frac{2.5}{22}\cdot\lambda \qquad [3.90]$$

For the same Nafion membrane, Zawodzinski *et al.* [ZAW 95] took a constant value of 1 for (ξ) for the water content range [GUP 08, BLU 07]. For ($\lambda < 5$), a proportionality relationship is adopted, with the authors considering that the electro-osmotic coefficient is ($\lambda = 0$). For ($\lambda > 14$), (ξ) increases linearly [ZAW 93]:

$$\xi = \begin{cases} 0.2\cdot\lambda & \text{for } \lambda>5 \\ 1 & \text{for } 5\leq\lambda\leq14 \\ 0.1875\cdot\lambda-1.625 & \text{for } \lambda>14 \end{cases} \qquad [3.91]$$

Moreover, this coefficient has recently been measured at around 1.07 in the studies [YE 07, LIU 07] for a Gore® membrane.

3.7. Conclusion

The equations describing the mass transfer in the different layers of a PEMFC are presented in this chapter. Stefan–Maxwell and Darcy models are useful for visualizing the diffusion of gaseous mixtures in the diffusion layers and the convective motion of the gas mixtures in the flux channels and the diffusion layers, respectively. The kinetic law of Butler–Volmer deals with the electrical representation of the electrochemical reactions that take place in the reaction layers.

The study of the mass transfer of different species involved in the fuel cell as well as the management of different flows are essential for its proper functioning, and consequently for a longer lifetime. Indeed, it should be supplied continuously with fuel (hydrogen) and oxidant (oxygen) to ensure the production of electricity. On the contrary, the products of the reaction at the heart of the fuel cell should also be continuously released as the mismanagement of the mass flows can lead to significantly lower performance in the fuel cell.

The flow of water in the membrane can describe the behavior of the fuel cell, which is why it contributes to the performance of a fuel cell. Two main descriptions (models) of this phenomenon have been described: porous medium model and phenomenological model. The transport of water in the membrane strongly depends on the overall thermoelectric behavior of the system. Maintaining a good membrane water content can guarantee good ion conductivity and consequently a good electrical performance of the PEMFC.

Heat Transfer Phenomena

4.1. Introduction

Understanding heat transfer phenomena within a fuel cell is essential in order to optimize its performance and contribute to its sustainability. The heat released from a PEMFC is relative to the electrical energy delivered by the battery itself. This generation of energy (heat) comes from the heat released in the irreversible electrochemical reaction (~55%), entropic heat (~35%) and the heat released by the Joule effect due to the ohmic resistance of the membrane (~10%).

In a PEMFC, heat is removed through a cooling system or transferred by conduction and convection at the cell interfaces.

Different modes of heat transfer are observed in the different components of the PEMFC. Indeed, heat transfer through the membrane is ensured by thermal conduction, while the conduction and convection are present at the same time in the catalytic layers (CL) and the diffusion layers (GDL). However, it should be noted that heat transfer and water transport in a PEMFC are always coupled: (i) the evaporation and condensation phenomena are respectively accompanied by the absorption and the latent heat release; (ii) the transport of water and heat are simultaneous (the temperature gradient induces the phase change and the transport of water); (iii) the pressure of the saturating vapor is strongly dependent on temperature [BOU 07, pp. 149–151].

Several models reflecting the heat transfer phenomenon in a PEMFC have emerged; initial efforts have been developed by Nguyen and White [NGU 93]. A PEMFC is built and regulated in such a way that its temperature is as uniform as possible to avoid creating hot spots in its core. The presence of these hot spots could have adverse effects on the materials, in particular on the polymer membrane. However, it is difficult to keep the temperature of the cell uniform, as a temperature gradient exists between the inlet and the outlet of the cooling circuit [COL 08]. This gradient could be used to boost the transport of water and thus the electrical performance of the battery. Strong thermal gradients, for example, would lead to the transport of strong water flows and this would limit waterlogging [BRA 06].

The simulation of these thermal phenomena by thermal models is therefore essential in order to be able to correctly estimate how a cell works when its temperature field is not uniform.

Thermal models have only recently appeared in the literature, with the exception of the article by Fuller and Newman [FUL 93] who were already interested in the thermal effects in the core of fuel cells and who stress the importance of coupled modeling of heat and mass transfers in the cell core to correctly describe the phenomenon of water sorption in the membrane and, consequently, its hydration. Nguyen and white [NGU 93] developed a two-dimensional (2D) model of a PEMFC in which they demonstrated a one-dimensional (1D) heat transfer model in the direction of flow. This model considers only the phase change of the water in the flow channels as the only heat source allowing convective heat transfer between the gas and the solid. Fuller and Newman [FUL 93] developed a pseudo-2D thermal model with a mass transfer (1D) through the membrane, and a heat transfer (1D) in the direction of flow.

More recently, the effects of temperature on the diffusion of water in the membrane have been modeled by Yan et al. [YAN 04].

The risks of membrane dehydration (in particular, at the anode) at high operating temperatures or at high current densities were shown through coupled modeling of heat and mass transfers in the membrane alone. Yi and Nguyen [YI 99] used Nguyen and White's model [NGU 93] to introduce the heat due to entropy and irreversibility at the same time as the heat due to phase change. This model allows the temperature of the solid phase to vary in the direction of flow only, assuming a uniform temperature in the direction across the membrane. Wöhr *et al.* [WÖH 98] developed a thermal model (1D) for heat and mass transfer across the plane direction of the PEMFC stack. While taking into account the entropic heat and that due to the irreversibility of the reaction, the authors of this work obtained the temperature profiles through the membrane and predicted the maximum temperature according to the number of cells contained in the stack. Rowe and Li [ROW 01] also developed a model (1D) taking into account the entropy, the heat released by the irreversibility of the reaction and the latent heat of phase change. They also took into account the heat released by the Joule effect in the membrane and at the electrodes. Since 2002, several studies have already been conducted to model heat transfer in a PEMFC.

Other studies agree on the importance of dealing simultaneously with heat transfers and material transfers in the stack. Thus, Djilali and Lu [DJI 02] presented the temperature profiles in the electrode membrane assembly (EMA) obtained from a one-dimensional model in a steady state.

The convective transport of reactive gases and water is taken into account, but the water remains in vapor form. Weber *et al.* [WEB 06] and Wang [WAN 06] set up coupled models of water and two-dimensional heat transfers taking into account changes in water conditions. The water flows in the GDL are not calculated by diffusion, but they are estimated from the streams of water vaporizing in the electrodes and condensing in the channels. These one-dimensional approaches allow a good understanding of the phenomena in the stack core, but the variations in concentration or temperature in the channels, resulting in a non-uniform distribution of the current densities in the cell, are not taken into account.

4.2. Energy balances for a PEMFC

4.2.1. *Energy balance for a stack*

The energy balance for a PEMFC stack is described as the sum of the incoming energies, which is equal to the sum of the outgoing energies. It is written as follows [BAR 05]:

$$\sum Q_{in} - \sum Q_{out} = W_{el} + Q_{dis} + Q_c \qquad [4.1]$$

where:

– Q_{in} is the enthalpy of the incoming reactive gases;

– Q_{out} is the enthalpy of the non-consumed gases and the heat produced by the reactants;

– W_{el} is the electricity generated;

– Q_{dis} is the heat dissipated;

– Q_c is the heat evacuated by the cooling system.

The heat is released by the electrochemical reaction that takes place inside the cell; it should be evacuated through a cooling system. The heat generated from a cell is associated with voltage losses. For a good estimate of the energy balance, it is necessary to take into account the energy contained in the reactants and the electricity produced [CHU 09]:

$$\frac{I}{nF} \cdot H_{PCS} \cdot n_{cell} = Q_{gen} + I \cdot V_{cell} \cdot n_{cell} \qquad [4.2]$$

with: $Q_{gen} = (1.254 - V_{cell}) \cdot I \cdot n_{cell}$

– Q_{gen} is the heat generated by the stack, in [W];

– n_{cell} is the number of cells;

– V_{cell} is the voltage of the cell, in [V].

If the produced water is in liquid form, the heat released by the stack is written as:

$$Q_{gen} = (1.482 - V_{cell}) \cdot I \cdot n_{cell} \qquad [4.3]$$

If the produced water is in the form of steam, the heat released by the stack is written as:

$$Q_{gen} = (1.254 - V_{cell}) \cdot I \cdot n_{cell} \qquad [4.4]$$

We can also write an equation based on the enthalpies, those entering will be the enthalpies of hydrogen, oxygen and water vapor present in these reactive gases.

The enthalpies of the outgoing flows are those of the electrical energy produced, the unconsumed reactive flows and the heat lost by radiation, convection or the cooling system. This new equation is written as:

$$\sum \left(h_i\right)_{in} = W_{el} + \sum \left(h_i\right)_{out} + Q \qquad [4.5]$$

The enthalpy of a dry gas, or a mixture of dry gas expressed in $[J. \ s^{-1}]$, is written as:

$$h = \dot{m} \cdot c_p \cdot T \qquad [4.6]$$

where:

– $\dot{m}$ is the mass flow rate of the gas or gas mixture, in $[g. \ s^{-1}]$;

– c_p is the specific heat, in $[J. \ g^{-1}.K^{-1}]$.

If we take the higher calorific value (HCV) of a gas, we write:

$$h = \dot{m} \cdot \left(c_p \cdot T + h_{PCS}^0\right) \qquad [4.7]$$

where h_{PCS}^0 is the enthalpy of the gas which corresponds to its HCV at 0°C, expressed in $[J. \ g^{-1}]$; calorific values are usually reported at 25°C, which means that the HCV needs to be calculated at the desired temperature [BAR 05, FEL 86]. The enthalpy of liquid water is written as:

$$h = \dot{m}_{H_2O(l)} \cdot c_{p,H_2O(l)} \cdot T \qquad [4.8]$$

The enthalpy of the water vapor is written as:

$$h = \dot{m}_{H_2O(g)} \cdot c_{p,H_2O(g)} \cdot T + h_{fg}^0 \qquad [4.9]$$

4.2.2. *Energy balance for compounds and gases*

The generalized energy balance equation for membrane electrode assembly (MEA) can be written in the form [FEL 86]:

$$Q_s = Q_c + h_f \cdot A_f \cdot \left(T_e - T_{f,av}\right) + h_a \cdot A_a \cdot \left(T_e - T_{a,av}\right) \qquad [4.10]$$

where:

– Q_c is the heat transmitted by conduction in the solid structure;

– Q_s is the heat source generated by the electrochemical reaction;

– h_f, h_a are the convective heat transfer coefficient of fuel H_2 (*f*) and air (*a*);

– A_f, A_a are the exchange surface with fuel H_2 (*f*) and air (*a*) respectively;

– T_e, $T_{f,av}$ and $T_{a,av}$ are the temperature of the electrode, the fuel and the air respectively.

4.2.3. *Energy balance for the gas phase*

Temperature gradients in the flow channels are generally assumed to depend solely on convection due to mass and heat transfer from the channel walls to the gas [SRI 06]:

$$\sum_i n_i \cdot c_p \cdot \frac{\partial T_g}{\partial x} + \sum_i c_p \cdot \frac{\partial n_i}{\partial x} \cdot \left(T_g - T_s\right) + h_g \cdot B_g \cdot \frac{1}{a} \cdot \left(T_g - T_s\right) = 0 \quad [4.11]$$

where:

– B_g is the ratio between the gas-heat exchange surface and the cell surface;

– T_g, T_s are the temperature of the gas and the solid;

– h_g is the convective transfer coefficient calculated from the Nusselt number, in $[W.m^{-2}.K^{-1}]$.

$$h_g = Nu \cdot \frac{k}{d_h} \qquad [4.12]$$

The Nusselt number depends on the channel geometry, the Reynolds number and the Prandtl number for a laminar flow.

4.2.4. *Energy balance for the solid structure*

The energy balance of the solid structure of the stack describes the conduction of heat, heat by convection (which is transmitted from the flow channels to the solid structure), and the reaction enthalpies occurring in the source terms [SRI 06]:

$$\frac{1}{s} \cdot \sum_g h_g \cdot B_g \cdot \left(T_g - T_s\right) - \frac{1}{s}\left(\frac{\Delta H}{ne \cdot F} + V\right) \cdot J + K_x \frac{\partial^2 T_s}{\partial x^2} = 0 \qquad [4.13]$$

where K_x being the effective thermal conductivity of the solid structure, with:

$$K_x = \frac{\sum_h K_h \cdot \delta_h}{\sum_h \delta_h} \qquad [4.14]$$

4.3. The heat flow in the different layers of the PEMFC

The heat flow density generated by the electrochemical reaction in the anodic catalytic layer is given by [CHE 05]:

$$q_{ACL} = j \cdot \left(\eta_a - \frac{\Delta H_a - \Delta G_a}{nF}\right) \qquad [4.15]$$

where:

- ΔH_a is the enthalpy of the reaction at the anode;

- ΔG_a is the Gibbs free energy at the anode.

This heat flow can be written as a function of the temperature gradient across the membrane:

$$q_{ACL} = -\lambda_{mem} \cdot \frac{dT}{dx} \qquad [4.16]$$

where λ_{mem} is the effective thermal conductivity of the membrane, in $[W.m^{-1}.K^{-1}]$.

4.3.1. *Heat transfer by conduction*

A temperature gradient in a homogeneous substance gives rise to a transfer of energy in this medium. The heat flow passing through a section (A) of this medium with a thermal conductivity, along the (Ox), is written as:

$$q_x = -k \cdot A \cdot \frac{dT}{dx} \qquad [4.17]$$

In the steady state, the heat transfer is governed by the equation:

$$\frac{d^2T}{dx^2} = 0 \qquad [4.18]$$

In the case of heat transfer through two adjacent walls with different thermal conductivities (two different materials), the boundary condition is such that the temperature at the adjacent interfaces is the same [BAR 05]:

$$q = h_{tc} \cdot A \cdot \Delta T \qquad [4.19]$$

where:

- h_{tc} is the convective heat transfer coefficient, in $[W.m^{-2}.K^{-1}]$;

- A is the surface in the direction of the heat flow, in $[m^2]$;

– ΔT is the temperature difference between the solid structure and the fluid.

The heat generated inside the stack can be described by Poisson's equation [INC 96]:

$$\frac{d^2T}{dx^2} + \frac{q_{int}}{k} = 0 \qquad [4.20]$$

where q_{int} is the amount of heat generated per unit of volume; it is considered as heat of electrical and ionic resistance and is written as:

$$q_{int} = j^2 \cdot \rho_s \qquad [4.21]$$

In a porous medium, an effective thermal conductivity that takes into account the porosity of the medium (ϵ) is used:

$$k_{eff} = -2k_s + \left[\frac{\epsilon}{2k_s + k} + \frac{1-\epsilon}{3k_s} \right]^{-1} \qquad [4.22]$$

4.3.2. Heat dissipation by natural convection and radiation

The heat flow lost by the stack by natural convection and radiation into the surrounding environment is written as [BAR 05]:

$$Q_{dis} = \frac{T_s - T_0}{R_{th}} \qquad [4.23]$$

where:

– T_s, T_0 are the temperature of the stack and the environment, respectively;

– R_{th} is the thermal resistance [m.K.W^{-1}], which is given by:

$$R_{th} = \frac{1}{\dfrac{1}{R_c} + \dfrac{1}{R_R}} \qquad [4.24]$$

Convective thermal resistance R_c is written as

$$R_c = \frac{1}{h \cdot A_s}$$ [4.25]

4.4. Thermal management in a PEMFC

For constant operation, the PEMFC must be maintained at a desired temperature. The continuous control of the PEMFC temperature is via the good thermal management of the different flows that exist in the fuel cell. The evacuation of these flows is essential, and a cooling system must therefore be considered.

4.4.1. *Cooling systems*

There are several methods to cool a PEMFC stack. The method chosen depends on the size and application of the stack. According to Shah [SHA 03], stacks whose power is < 100 W can be cooled only by the air that enters the cathode side (considered as a reactant); stacks whose power varies between 200 W and 2 kW can be cooled by channels separate to those of the reactants. Stacks whose power is > 10 kW require a cooling liquid. Faghri [FAG 05] briefly described the problems of thermal transfer in a PEMFC. Faghri [FAG 08] presented the possibility of integrating a heat pump into the stack in order to improve its thermal management. In one of his designs, he proposes to integrate a heat pump in the lower part of the bipolar plates. Reichler [REI 09] presented a theoretical study on cooling performance through innovative cooling systems for fuel cell electric vehicles.

4.4.2. *Convection cooling of the airflow at the cathode*

This is one of the simplest solutions; this method does not require a complicated architecture or an industrial cooling fluid. This system (see Figure 4.1) is desirable for small stacks delivering a low power (<100 W). The heat dissipation is ensured by the flow of air, also considered as a reactant, in the flow channels on the cathode side. Moreover, this method does not control the stack temperature; it depends solely on the temperature and humidity of the surrounding air, hence why the regulation of the stack temperature is not ensured by this method of cooling. The latter also requires larger flow channels on the cathodic side compared to those on the anode side, and consequently the stack volume becomes larger.

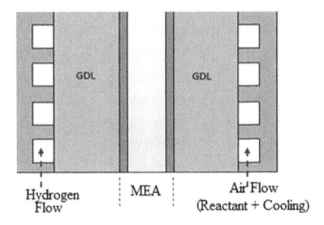

Figure 4.1. *Convection cooling of the airflow at the cathode*

4.4.2.1. *Cooling with separate air channels*

This type of cooling is recommended for a stack with a power of a few hundred watts. Figure 4.2 shows a diagram of this cooling method.

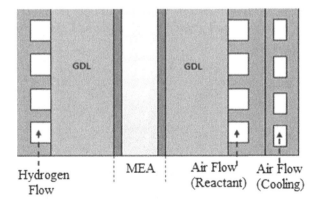

Figure 4.2. *Cooling with separate air channels*

4.4.2.2. *Cooling using a liquid*

The thermal properties (specific heat, thermal conductivity) of a liquid are higher than the air or even another gas. These coolants, used in separate channels in the bipolar plate, are used in medium and high power PEMFCs for automotive applications (see Figure 4.3).

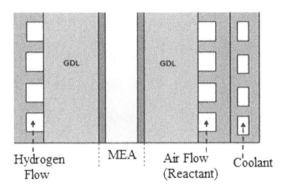

Figure 4.3. *Cooling using a liquid coolant*

4.4.2.3. *Cooling by evaporation*

The enthalpy of vaporization of water is used in the cooling by injecting an additional quantity of water into the reactant flows. The heat generated at the electrodes is used (consumed) by the water to change its state by keeping the temperature constant (see Figure 4.4). This method also helps prevent the membrane from drying.

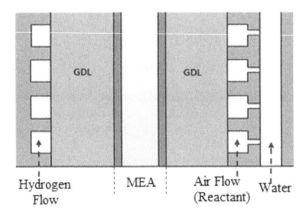

Figure 4.4. *Cooling by evaporation*

4.4.2.4. *Integrated heat sink cooling*

This method is obviously ingenious; it involves the heat sink phenomenon as it provides efficient heat transfer from the stack by thermal conduction through the sink, then by convection (natural or forced) to the surrounding environment (see Figure 4.5).

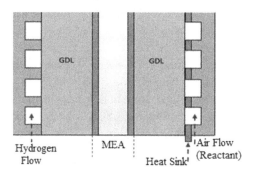

Figure 4.5. *Integrated heat sink cooling*

4.4.2.5. *Cooling by plates*

This is the solution that is found most often in the PEMFC stack; these plates are prefabricated and inserted into the stack. There are even perforated bipolar plates designed specifically to carry coolants (air, water, etc.), in addition to hydrogen and oxygen. For this method, a whole control system should be considered; temperature regulation would be essential to avoid the dehydration of the membrane or its blockage with excess water. Figure 4.6 shows a typical diagram of a plate cooling system.

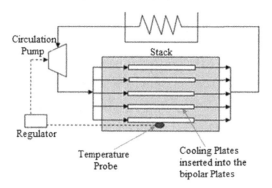

Figure 4.6. *Cooling by plates*

4.4.3. *The effect of temperature on the performance of the PEMFC*

The temperature is of paramount importance, directly or indirectly influencing the performance of a PEMFC. Indeed, variations in temperature affect phenomena related to the reaction kinetics, the transport of water, the humidity, the membrane conductivity, the tolerance of the catalyst and the dissipation of heat. A good review of the role of temperature in PEMFCs was presented by Zhang *et al.* [ZHA 06] also aimed at the development of high temperature PEMFCs. Song *et al.* [SON 07] experimentally studied the performance of the Nafion membrane [TAN 97], which constitutes a PEMFC. The authors concluded that battery performance increases with increasing temperature from ambient to 80°C. They also found that at low current density (< 0.4 A/cm^2), an increase in temperature decreases the performance of the fuel cells; for high current densities (> 0.4 A/cm^2), an increase in temperature implies an increase in performance. In their experiments, they found that the best performance was at around 80°C, with an absolute pressure of 3 bar and a relative humidity of 100%.

4.5. Heat sources in the PEMFC

Wang [WAN 04] mentioned in his review of PEMFC modeling that, in most modeling studies, the heat released due to entropy change has always been neglected. There is a good deal of fuel cell modeling going from simple empirical models to complex CFD models. The existence of journals in the literature such as [BLY 05, SIE 08, WEB 04a] clearly indicates the extent of ongoing research around the world on the subject of PEMFCs. In work by Hashmi [HAS 10], it is well explained that, in the literature, there is no consensus on the magnitude and location of the reversible heat production at each electrode. Several modeling studies (non-isothermal) of fuel cells [BAC 08, NGU 04, SAD 09] have been based on Lampinen results [LAM 93] for the entropy change if all the reversible heat is produced at the cathode.

Wöhr *et al.* [WÖH 98] used different values for entropy change at the anode and at the cathode, except that the change in total entropy (for the whole reaction) did not correspond to the change in entropy of the two half reactions.

Therefore, the results show irregular heat generation at the electrodes. Ramousse *et al.* [RAM 09] suggested that it is due to the hydrogen oxidation reaction being highly exothermic, while the oxygen reduction reaction is endothermic.

There are few reviews in the literature on the topic of cooling channels in PEM fuel cells. On the contrary, many studies are underway on the transport of heat and water in this same type of fuel cell and for its different layers.

Chen *et al.* [CHE 03] numerically studied different types of cooling methods (parallel and serpentine channels). They used a criterion known as Index of Uniform Temperature (IUT). In their simulations, they used the uniform heat flow generated by the stack.

At the inlet of the channel, they specified the mass flow and the flow velocity values, but they did not mention anything for the channel outlet.

They assumed that the surrounding surface at the cooling plates is adiabatic, and that the surrounding area along the edge of the flat plate is small.

For the heat transfer from the solid structure to the fluid, the convective heat transfer coefficient is estimated from a correlation taking into account the conjugated heat transfer.

On conclusion of their work, they indeed found that cooling with serpentine channels was better than parallel channels with consequent charge loss.

Choi *et al.* [CHO 08] digitally analyzed the performance of the plate cooling system in the PEM fuel cell. These studies are similar to those of Chen *et al.* [CHE 03]. They studied the two types of panel configurations (parallel and serpentine) using maximum surface temperature as an optimization criterion. They also used a uniform heat flux for heat generation and analyzed for three different values of heat flow, the case where there is one, three and finally five cell units in a stack. From their results, they concluded that the maximum surface temperature increases linearly, the heat flux increases for their two particular designs with two different Reynolds numbers.

Yu *et al.* [YU 09] assessed the design of four Multi-Pass Serpentine Flow-Fields (MPSFF) with two configurations: coil or conventional spiral for the cooling plates of PEM fuel cells. They used the IUT criterion to compare configurations, as previously defined by Chen *et al.* [CHE 03]. They compared the configurations with the maximum surface temperature and with the difference in surface temperature (maximum and minimum).

With regard to the geometry, they preferred symmetrical geometries to reduce the number of nodes in their simulations, thereby limiting the use of the same heat flow values on both sides of the cooling plate.

Three different uniform heat flux values are used in the simulation to cover the normal operating range of the PEM fuel cell. The velocity is introduced as a boundary condition at the entrance of the channel. Their results suggest that the IUT decreases and the pressure loss increases as the Reynolds number increases in the six configurations.

The MPSFF design is motivated by the study of Xu *et al.* [XU 07] for flow field structures of PEM fuel cell reactants, which is based on the reforming of a single serpentine flow field, such that the pressure difference between the adjacent flow channels improves the flow through the porous electrode, as an interdigital flow field.

First, it is necessary to know the spatial distribution and the intensity of the heat sources. There are three types of heat produced in a PEMFC in operation: the heat due to the electrochemical reaction, the heat due to the passage of the current (Joule effect) and the heat due to water phase changes [COL 08].

4.5.1. *In the polymer membrane*

At the membrane level, the Joule effect is caused by proton transfer resistance. It should be noted that the calculated resistance is that of the membrane as a whole. The variations in conductivity in the thickness of the membrane are neglected, and the heat source associated with the Joule effect is uniformly distributed in its thickness. This volume density of heat flow ($\dot{q}_J$), expressed in [$W.m^{-3}$] [RAM 09], is written as:

$$\dot{q}_J = \frac{R_m \cdot j^2}{e_m} \qquad [4.26]$$

where:

– R_m is the electrical resistance of the membrane, in $[\Omega.cm^2]$;

– e_m is the thickness of the membrane, in $[cm]$.

4.5.2. *At the electrodes*

The hydrogen oxidation and oxygen reduction reactions at the cathode are exothermic and the production of heat is assumed to be homogeneous. The water is produced by the reaction agglomerates and it is assumed to be in liquid form. However, three heat sources are observed at the electrodes:

– heat flow (reversible) released by the electrochemical reaction (variation in entropy of each half-reaction);

– heat flow (irreversible) of the electrochemical activation of reactions;

– phase change heat flow (sorption).

For the anode, we write:

$$\dot{Q}_a = \dot{Q}_a^{reac} + \dot{Q}_a^{act} + \dot{Q}_a^{sorp} \qquad [4.27]$$

and for the cathode :

$$\dot{Q}_c = \dot{Q}_c^{reac} + \dot{Q}_c^{act} + \dot{Q}_c^{sorp} \qquad [4.28]$$

4.5.2.1. *Heat flow of the reaction*

The total heat released by the reaction at the two electrodes Q^{reac} in $[J.mol^{-1}]$ is expressed according to the variation in entropy ΔS^{rev} [ROS 99], such that:

$$Q^{reac} = -T \cdot \Delta S^{rev} \qquad [4.29]$$

Sørensen and Kjelstrup [SOR 02] indicated that $\Delta S^{rev} = 163.12$ [J.mol^{-1}.K^{-1}], for T = 353 K. The corresponding heat flow $\dot{Q}^{reac}$ expressed in [kW.m^{-2}] is written for a current density, i, and for water in the liquid form:

– for the heat flow at the anode, we write:

$$\dot{Q}_a^{reac} = \frac{j}{2F}.Q_a^{reac} \tag{4.30}$$

– for the cathode:

$$\dot{Q}_c^{reac} = \frac{j}{2F}.Q_c^{reac} \tag{4.31}$$

The two electrodes are therefore sources of heat. However, few relevant values exist for the half-reaction occurring at the anode, the authors of [KJE 03, WU 07] believed that the reaction is athermic, that is, the heat is considered to be produced at the cathode. According to their results, we will only have to consider the heat flow for the cathode:

$$Q_a^{act} = -48.7 \frac{j}{2F} \tag{4.32}$$

4.5.2.2. Heat flow by electrochemical activation

The irreversibility of electrochemical reactions results in overvoltages at the electrodes [ROS 03]. According to the theory of the activated complex, these electrochemical reactions are responsible for the degradation of part of the energy created by the reaction [RAM 09]. The writing of the Butler–Volmer law, modeling the reaction activation phenomena or the modeling of the coupled mass and charge transfers at the electrode, makes it possible to estimate the corresponding heat sources. These quantities of heat are written according to the variation in entropy, such that:

$$Q_a^{act} = -T \cdot \Delta S_a^{irr} \tag{4.33}$$

and:

$$Q_c^{act} = -T \cdot \Delta S_c^{irr} \tag{4.34}$$

Note that these irreversibilities are responsible for the anode A and cathode C overvoltages, respectively; they are therefore estimated from the transport model. The heat flow released by these overvoltages is calculated as in [RAM 09]:

$$\dot{Q}_a^{act} = |\eta_A| . j \qquad [4.35]$$

and:

$$\dot{Q}_c^{act} = |\eta_C| . j \qquad [4.36]$$

The overvoltage at the anode is assumed to be negligible compared to that at the cathode, that is, $\dot{Q}_c^{act} = 0$. As for cathode overvoltage, it depends on the amount of water in the electrodes [COL 08].

4.5.3. *In the GDLs*

For the heat released by phase changes, the water is both in vapor form (in GDL) and in liquid form in the membrane; it can change state within the fuel cell. Two phase change mechanisms are to be considered: the phase change in a GDL and the phase change at the interfaces with the polymer membrane.

In fact, the water in the membrane and the water produced are in liquid form. Water entering the membrane condenses, causing heat to escape. When it comes out, it must vaporize (if in the pores of the electrodes, the vapor pressure is below the saturation vapor pressure, the water produced or that which exits the membrane vaporizes), creating a heat sink. Knowing the water flow through the membrane and the enthalpy of water sorption in the membrane, respectively $N_{H_2O}^m$ and $\Delta H_{H_2O}^{sorp}$, is enough to determine the intensity of the corresponding heat sources or sinks. It is therefore necessary to take into account the quantity of heat produced or consumed during sorption (adsorption or desorption) of the water on the surface of the membrane and the desorption of the water produced in the electrode.

The sorption phenomena thus reflect the equilibrium between a liquid phase (adsorbed) and a gaseous phase (desorbed) at an interface. Transfers are therefore subject to a phase change at this interface.

Several authors have focused on measuring the enthalpy of water sorption in a Nafion® polymer membrane. Watari *et al.* [WAT 03] reviewed the values used in the literature and reported values varying between 42 and 52 kJ.mol^{-1} according to the water content of the membrane or its pretreatment. However, there was no clear correlation between measured values and experimental conditions. More recently, Burnett *et al.* [BUR 06] took measurements using Nafion®112 and obtained values between 43 and 60 kJ.mol^{-1}.

It is not currently possible to know the exact sorption enthalpy as a function of the operating conditions. However, the measured values are close to the latent heat of water vaporization L_{vap} (Atkins and De Paula [ATK 02] give $L_{vap} = 41.6$ kJ.mol^{-1} to 353 K). In what follows, the sorption enthalpy of water in the membrane will be considered equal to the latent heat of vaporization of the water L_{vap} [ROS 99].

According to Ramousse *et al.* [RAM 09], the algebraic character of the calculated water flows makes it possible to treat sorption and desorption in the same way. The heat flow densities are expressed in [W.m^{-2}]:

$$\dot{Q}_a^{sorp} = \Delta H_{H_2O}^{sorp} \cdot N_{H_2O}^a \qquad [4.37]$$

and:

$$\dot{Q}_c^{sorp} = -\Delta H_{H_2O}^{sorp} \cdot N_{H_2O}^c \qquad [4.38]$$

In order for the production of water in liquid form at the cathode to be generated, we write:

$$N_{H_2O}^c = N_{H_2O}^a + \frac{j}{2F} \qquad [4.39]$$

4.5.4. *Water evaporation and condensation*

When the water is in liquid form in part of the GDL and in the vapor form in the other, at the interface between these two zones, the water changes state and there will be production (or absorption) of the heat [ROS 99]. The amount of heat required for vaporization or that released by the condensation of a stream of water is estimated from the latent heat of vaporization such as:

$$\dot{Q}^{cond} = L_{vap}.N_{H_2O}^{vap \rightarrow liq} \qquad [4.40]$$

and for the vaporization:

$$\dot{Q}^{vapo} = -L_{vap}.N_{H_2O}^{liq \rightarrow vap} \qquad [4.41]$$

where:

– $N_{H_2O}^{vap \rightarrow liq}$ is the molar flow density of condensing steam, in $[mol.m^{-2}.s^{-1}]$;

– $N_{H_2O}^{liq \rightarrow vap}$ is the molar flow density of vaporizing liquid water, in $[mol.m^{-2}.s^{-1}]$.

These water flow densities are estimated from the water model found in the literature. In addition, the passage of charged particles in a medium causes, as already mentioned, the release of heat due to the Joule effect. GDL and electrodes are made of carbon materials and they are good electronic conductors; their electrical resistance is low and we neglect the heat generated by the Joule effect due to the passage of electrons in the electrodes, the GDL and also in the bipolar plates.

The electrical contact of the different layers is assumed to be perfect. In summary, three heat sources (or sinks) coexist within the stack. The first is localized at the electrodes and corresponds to the heats of reaction, overvoltages and adsorption/desorption. The second is point sources of heat in GDLs when there is a phase change of water. Finally, the third is a "volume" heat source located in the membrane.

Figure 4.7 shows the location of the heat sources within a PEFC. In GDLs, water flows are usually oriented from the electrode to the GDL.

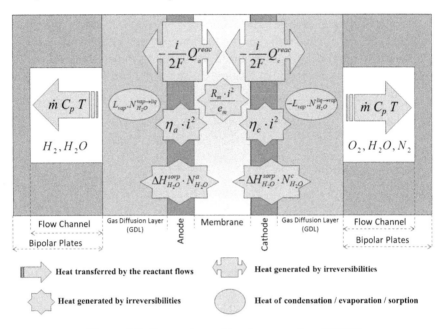

Figure 4.7. *Expressions of heat sources in a PEMFC*

To illustrate everything in this figure, the water changing state within the GDL and the heat necessary for the vaporization of the liquid water are taken into account.

To determine the thermal energy produced in a different way, it is necessary to remove the electrical energy from the total chemical energy available for the reaction. In addition, the water present in the stack can be in liquid form or in vapor form. It is necessary to take into account the heat released or absorbed within the fuel cell of during phase changes. The total heat produced by the fuel cell depends on the operating point. It is given by:

$$Q_{total} = \left(\frac{\Delta h(T,P)}{2F} - E(T,P) \right) \cdot J + Q_{H_2O}$$ [4.42]

where:

– $\Delta h(T,P)$ is the enthalpy of the reaction when water is produced in liquid form;

– $E(T,P)$ is the voltage at the terminals of the fuel cell (function of J);

– Q_{H_2O} is the amount of heat resulting from the phase changes of water;

– J is the current delivered, in [A].

4.6. Temperature distribution between two cathodes: case study

In this section, we propose to study the temperature distribution at the interfaces of the components of an elementary PEMFC presented in Figure 4.8.

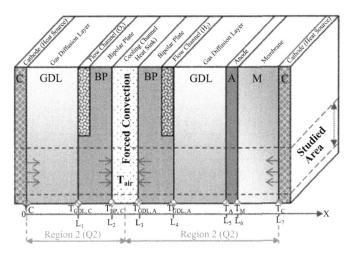

Figure 4.8. *The elementary cell studied*

The domain to be studied consists of several interfaces shared across two regions separated by a cooling channel. The heat highlighted in this model is that released by the irreversible electrochemical reaction and the entropic heat (~90% of the total heat released by a cell). The heat generated by the Joule effect, due to the ohmic resistance of the membrane (~10%), will not be considered in this case study. This heat is located at the cathode of each heat source region; it will be transferred by thermal conduction through the different layers of the cell to the heat sink (cooling channel).

The heat flow density to be extracted by the cooling channel is the sum of the two heat flow densities from the two regions considered; it can be expressed as follows: the thermal convection coefficient (h) is determined as a function of the geometry of the cooling channel and the flow type. Table 4.1 summarizes the numerical values used for this simulation:

$$Q_{ext} = Q_1 + Q_2 \qquad\qquad [4.43]$$

Parameters	Numerical values
$e_{m/m}$	183 µm/0.36
$e_{A/A}$	20 µm/0.8
$e_{GDL/GDL}$	250 µm/1.25
$e_{BP/BP}$	6 mm/15
Tair	30°C
Tc	80°C
$T_{moy} = T_{BP} + T_{air}/2$	54°C
a/b (cooling channel)	6/2
Reynolds Number Re (at $3m.s^{-1}$)	494
Prandtl Number Pr	0.7
Nusselt Number Nu (internal laminar flow)	3.96
h_{air}	35

Table 4.1. *Numerical values of the simulation*

with (Q), the amount of heat extracted by the cooling channel or released by the two regions, expressed in [W]. It is assumed that the temperature of the cathode is uniform and constant (Dirichlet condition); the thermal balance of the region (1) is written as:

$$Q_1 = \frac{T_C - T_{BP,A}}{\dfrac{e_m}{\lambda_m} + \dfrac{e_A}{\lambda_A} + \dfrac{e_{GDL}}{\lambda_{GDL}} + \dfrac{e_{BP}}{\lambda_{BP}}} = h_{air} \times \left(T_{BP,A} - T_{air} \right) \qquad [4.44]$$

where:

– T_i is the temperature at the interface (i), in [K];

– e_i is the width of interface (i), in [m];

– h is the thermal convection coefficient, in $[W.m^{-2}.K^{-1}]$;

– λ is the thermal conductivity, in $[W.m^{-1}.K^{-1}]$.

Similarly for region (2):

$$Q_2 = \frac{T_C - T_{BP,C}}{\dfrac{e_{GDL}}{\lambda_{GDL}} + \dfrac{e_{BP}}{\lambda_{BP}}} = h_{air} \times \left(T_{BP,C} - T_{air} \right) \qquad [4.45]$$

The temperature distribution is obtained from Fourier's law:

$$\rho c \frac{\partial T}{\partial t} = \frac{\partial}{\partial x}\left(\lambda \frac{\partial T}{\partial x} \right) + \frac{\partial}{\partial y}\left(\lambda \frac{\partial T}{\partial y} \right) + \frac{\partial}{\partial z}\left(\lambda \frac{\partial T}{\partial z} \right) + S \qquad [4.46]$$

For one-dimensional heat transfer in a steady state and without a source term, we have:

$$\frac{\partial^2 T}{\partial x^2} = 0 \qquad [4.47]$$

The solution of this equation is in the form:

$$T(x) = c_1 \cdot x + c_2 \qquad [4.48]$$

Constants C1 and C2 are determined from the boundary conditions presented in Table 4.2. The thermal resistances are grouped in Table 4.3.

Interfaces	Boundary conditions
$x = 0$	$T(0) = T_C$
$x = L_1$	$T(L_1) = T_{GDL,C}$
$x = L_2$	$T(L_2) = T_{BP,C}$
$x = L_3$	$T(L_3) = T_{BP,A}$
$x = L_4$	$T(L_4) = T_{GDL,A}$
$x = L_5$	$T(L_5) = T_A$
$x = L_6$	$T(L_6) = T_m$
$x = L_7$	$T(L_7) = T_C$

Table 4.2. *Boundary conditions of the model*

Elements	Expressions
Membrane (m)	$R_m = e_m / \lambda_m$
Anode	$R_A = e_A / \lambda_A$
Diffusion Layer GDL	$R_{GDL} = e_{GDL} / \lambda_{GDL}$
Bipolar Plate BP	$R_{BP} = e_{BP} / \lambda_{BP}$
Region (1)	$R_1 = R_m + R_A + R_{GDL} + R_{BP}$
Region (2)	$R_2 = R_{GDL} + R_{BP}$

Table 4.3. *Thermal resistances of different elements of the model*

The equations for interfacial temperatures are summarized in Table 4.4.

Interfaces	Temperature
$x=0$	T_C
$x=L_1$	$T_{GDL,C} = \left(T_C + \left(\left(R_{GDL}/R_{BP}\right) \cdot T_{BP,C}\right)\right) / \left(\left(R_{GDL}/R_{BP}\right) + 1\right)$
$x=L_2$	$T_{BP,C} = \left(T_C + \left(h_{air} \cdot R_2 \cdot T_{air}\right)\right) / \left(\left(h_{air} \cdot R_2\right) + 1\right)$
$x=L_3$	$T_{BP,A} = \left(T_C + \left(h_{air} \cdot R_1 \cdot T_{air}\right)\right) / \left(\left(h_{air} \cdot R_1\right) + 1\right)$
$x=L_4$	$T_{GDL,A} = \left(h_{air} \cdot R_{BP} \cdot \left(T_{BP,A} - T_{air}\right)\right) + T_{BP,A}$
$x=L_5$	$T_A = \left(\left(R_{GDL}/R_{BP}\right) \cdot \left(T_{GDL,A} - T_{BP,A}\right)\right) + T_{GDL,A}$
$x=L_6$	$T_m = \left(T_C + \left(\left(R_m/R_A\right) \cdot T_A\right)\right) / \left(\left(R_m/R_A\right) + 1\right)$
$x=L_7$	T_C

Table 4.4. *Summary of interfacial temperatures*

Equations that give the temperature profile for each cell element are shown in Table 4.5.

Distances	Temperature distribution
$0 \rightarrow L_1$	$T(x) = \left[\left(\left(T_{GDL,C} - T_c\right)/e_{GDL}\right) \cdot x\right] + T_c$
$L_1 \rightarrow L_2$	$T(x) = \left[\left(\left(T_{BP,C} - T_{GDL,C}\right)/e_{BP}\right) \cdot x\right] + T_{GDL,C}$
$L_7 \rightarrow L_6$	$T(x) = \left[\left(\left(T_m - T_c\right)/e_m\right) \cdot x\right] + T_c$
$L_6 \rightarrow L_5$	$T(x) = \left[\left(\left(T_A - T_M\right)/e_A\right) \cdot x\right] + T_m$
$L_5 \rightarrow L_4$	$T(x) = \left[\left(\left(T_{GDL,A} - T_A\right)/e_{GDL}\right) \cdot x\right] + T_A$
$L_4 \rightarrow L_3$	$T(x) = \left[\left(\left(T_{BP,A} - T_{GDL,A}\right)/e_{BP}\right) \cdot x\right] + T_{GDL,A}$

Table 4.5. *Summary of the equations giving the temperature profiles*

The temperature distribution in the cell is shown in Figure 4.9.

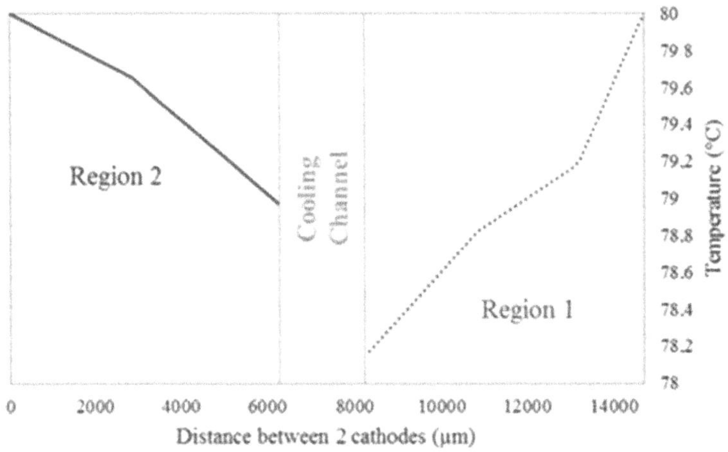

Figure 4.9. *Temperature distribution in the cell*

It can be seen that the difference in temperature between the heat source (the cathode) and the heat sink (the cooling channel) cannot exceed 1.5°C for both regions, which makes us consider the cell to be isothermal.

4.7. Conclusion

Modeling heat transfers in the cell is essential for a detailed analysis of the performances of a fuel cell. The dependence of mass and charge transfer phenomena on temperature shows the importance of an evaluation of the temperature field in the fuel cell core.

The thermal model developed in this chapter allowed us to evaluate a temperature difference (of about 1.5°C) in an area between two cathodes. This result is the subject of a recommendation in work by Chupin [CHU 09]. This gradient should not increase as drying (or even deterioration) of the membrane could cause the fuel cell to stop operating. It is therefore necessary to provide good thermal management of the fuel cell.

In general, the heat dissipation results in heating of the gases and the cooling of water in the bipolar plate. Thus, global modeling allows the coupling to modeling of mass and charge transfer along the supply channels. This modeling can also provide information on the risk of saturation. There may then be a risk of water condensation, which may restrict the access of gases to the electrodes, thus causing a sharp drop in the performance of the fuel cell.

The consumption of reactants and the evacuation of heat and water in the flow channels are responsible for the significant heterogeneous distribution of current densities in the cell.

In our case, which is just given as an example, the heat produced is evacuated by the circulation of air between the bipolar plates. It can, by a different technique, be evacuated through the flow of supply gas and water in cooling channels in the bipolar plate. It would then be interesting to evaluate the heating of the gases and water between the inlet and the outlet of the bipolar plate in order to be able to describe the temperature variations in the cell.

List of Symbols

Nomenclature

Symbol	Meaning	Unit
a	Water activity	$-$
A	Surface area	$[cm^2]$
A_{ac}	Empirical coefficient	[V]
b	Empirical coefficient	$[A.cm^{-2}]$
c	Molar concentration	$[mol.\ m^{-3}]$
C_p	Heat capacity	$[J.\ kg^{-1}.K^{-1}]$
D	Diffusion coefficient	$[m^2.s^{-1}]$
E	Reversible voltage	[V]
EW	Equivalent weight	$[g.mol^{-1}]$
F	Faraday constant	$[C.\ mol^{-1}]$
h	Convection coefficient	$[W.m^{-2}.K^{-1}]$
J	Current	[A]
j	Current density	$[A.cm^{-2}]$
j_0	Exchange current density	$[A.cm^{-2}]$
k	Heat conductivity	$[W.m^{-1}.K^{-1}]$
L	Length	$[\mu m]$
m	Empirical coefficient	[V]

n	Empirical coefficient	$[cm^2.A^{-1}]$
n_e	Number of electrons transferred	–
P	Pressure	[atm]
Q	Heat quantity	[J]
$\dot{Q}$	Heat flow	[W]
Q	Electric charge	[Coulombs. mol^{-1}]
$\dot{q}$	Heat flow density	$[W /m^2]$
R	Ideal gas constant	$[J\ K^{-1}\ mol^{-1}]$
r	Specific resistance	$[_\Omega .cm^2]$
T	Temperature	[K]
V	Voltage	[V]
Vcell	Cell voltage	[V]
We	Electrical work	[J]
x	Mass fraction	–

Notations associated with the Greek alphabet

Symbol	Meaning	Unit
α	Charge transfer coefficient	–
ΔH	Variation in enthalpy	$[J.\ mol^{-1}]$
ΔG	Gibbs free energy	$[J.\ mol^{-1}]$
η	Voltage loss	[V]
ζ_{air}	Stoichiometry of air	–
λ	Water content	–
σ	Membrane conductivity	$[S.cm^{-1}]$
$\digamma$	Density	$[Kg.m^{-3}]$

Indices and exponents

Symbol	Meaning	–
a	Anode	–
ac	Anode Cathode	–
ACL	Anode Catalytic Layer	–
act	Activation	–
c	Cathode	–
conc	Concentration	–
H^+	Proton	–
m	Membrane	–
max	Maximum (current)	–
ohm	Ohmic	–
Sorp	Sorption	–
sys	System	–
th	Theoretical	–

Bibliography

[AFH 18] AFHYPAC, available at: http://www.afhypac.org/documents/tout-savoir/ Fiche%203.2.1%20-%20Electrolyse%20de%20l%27eau%20rev%20%20Janv.%20 2018-2%20ThA.pdf, visited 13 February 2018.

[AMA 03] AMARAL SOUTO H.P., Diffusion-dispersion en milieu poreux: étude numérique de tenseur de dispersion pour quelques arrangements périodiques bidimensionnels ordonnés et désordonnés, PhD thesis, LEMTA – Institut national polytechnique de Lorraine, 2003.

[AMP 95a] AMPHLETT J.C., BAUMERT R.M., MANN R.F. *et al.*, "Performance modeling of the BallardMark IV solid polymer electrolyte fuel cell I. Mechanistic model development", *Journal of the Electrochemical Society*, vol. 142, pp. 1–8, 1995.

[AMP 95b] AMPHLETT J.C., BAUMERT R.M., MANN R.F. *et al.*, "Performance modeling of the ballard mark IV solid polymer electrolyte fuel cell II. Empirical model development", *Journal of the Electrochemical Society*, vol. 142, pp. 9–15, 1995.

[AMU 03] AMUNDSON R.N., PAN T.W., PAULSEN V.I., "Diffusing with Stefan Maxwell", *AIChE Journal*, vol. 49, no. 4, pp. 813–830, 2003.

[APP 88] APPLEBY A., FOULKES F.R., *Fuel Cell Handbook*, Van Nostrand Reinhold, New York, 1988.

[ATK 02] ATKINS P., DE PAULA J., *Physical Chemistry*, Oxford University Press, Oxford, 2002.

[AUP 13] AUPRETRE F., L'hydrogène comme vecteur énergétique, Compagnie européenne des technologies de l'hydrogène, available at: http://ipnweb.in2p3.fr/ ~ed421/ED421/hydrogene.pdf, visited 18 August 2013.

[BAC 08] BACA C.M., TRAVIS R., BANG M., "Three-dimensional single phase non isothermal CFD model of a PEM fuel cell", *Journal of Power Sources*, vol. 178, pp. 269–281, 2008.

[BAL 02] BALKIN A.R., Modeling a 500 W Polymer Electrolyte Membrane Fuel Cell, Project, University of Technology, Sydney, 2002.

[BAR 01] BARD J.A., FAULKNER L.R., *Electrochemical Methods: Fundamentals and Applications*, 2nd ed., John Wiley & Sons, New York, 2001.

[BAR 05] BARBIR F., *PEM Fuel Cells: Theory and Practice*, Elsevier Academic Press, Oxford, 2005.

[BER 68] BERGER C., *Handbook of Fuel Cell Technology*, Prentice Hall, New York, 1968.

[BER 91] BERNARDI D.M., VERBRUGGE M.W., "Mathematical model of a gas diffusion electrode bonded to a polymer electrolyte", *AIChE Journal*, vol. 37, no. 8, pp. 1151–1163, 1991.

[BER 92] BERNARDI D.M., VERBRUGGE M.W., "A mathematical model of the solid polymer electrolyte fuel cell", *Journal of the Electrochemical Society*, vol. 139, no. 9, pp. 2477–2491, 1992.

[BER 02] BERNING T., LU D.M., DJILALI N., "Three dimensionnal computational analysis of transport phenomena in a PEM fuel cell", *Journal of Power Sources*, vol. 106, pp. 284–294, 2002.

[BIR 02] BIRD R.B., STEWART W.E., LIGHTFOOT E.N., *Transport Phenomena*, 2nd ed., John Wiley & Sons, New York, 2002.

[BLU 07] BLUNIER B., MIRAOUI A., *Piles à Combustible: principes, modélisation, applications avec exercices et problèmes corrigés*, Éditions Ellipses, Paris, 2007.

[BLY 05] BLYLKOGLU A., "Review of Proton exchange membrane fuel cell models", *International Journal of Hydrogen Energy*, vol. 30, pp. 1181–1212, 2005.

[BON 94] BONTHA J.R., PINTAURO P.N., "Water orientation and ion solvation effects during multicomponentsalt partitioning in a Nafion cation exchange membrane", *Chemical Engineering Science*, vol. 49, no. 23, pp. 3835–3851, 1994.

[BOO 51] BOOTH F.J., "The dielectric constant of water and saturation effect", *Journal of Chemical Physics*, vol. 19, pp. 391–394, 1951.

[BOU 03] BOUSQUET S., Étude d'un système autonome de production d'énergie couplant un champ photovoltaïque, un électrolyseur et une pile à combustible: réalisation d'un banc d'essai et modélisation, PhD thesis, École des mines de Paris, available at: http://tel.archives-ouvertes.fr/docs/00/50/00/25/PDF/these_SBusquet.pdf, 2003.

[BOU 07] BOUDELLAL M., *La pile à combustible: structure, fonctionnement, applications*, Dunod, Paris, 2007.

[BRA 06] BRADEAN R., HAAS H., EGGEN K. *et al.*, "Stack models and designs for improving fuel cell startup from freezing temperatures", *ECS Transactions*, vol. 3, no. 1, pp. 1159–1168, 2006.

[BRA 07] BRANDELL D., KARO J., LIIVAT A. *et al.*, "Molecular dynamics studies of the Nafion, Dow and Aciplex fuel cell polymer membrane systems", *Journal of Molecular Modeling*, vol. 13, no. 10, pp. 1039–1046, 2007.

[BRU 99] BRUCE L., Conceptual design and modeling of a fuel cell scooter for urban Asia, Master's thesis, Princeton University, 1999.

[BUR 06] BURNETT D.J., GARCIA A.R., THIELMANN F., "Measuring moisture sorption and diffusion kinetics on proton exchange membranes using a gravimetric vapor sorption apparatus", *Journal of Power Sources*, vol. 160, pp. 426–430, 2006.

[CEA 18] http://www.cea.fr/Documents/Les%20technologies%20de%20l%E2%80%99hydrog%C3%A8ne%20au%20CEA.pdf, accessed on May 31st, 2018.

[CHE 03] CHEN F.C., GAO Z., LOUTFY R.O. *et al.*, "Analysis of optimal heat transfer in a PEM fuel cell cooling plate", *Fuel Cells*, vol. 3, pp. 181–188, 2003.

[CHE 05] CHEN R., ZHAO T.S., "Mathematical Modeling of a passive feed DMFC with heat transfer effect", *Journal of Power Sources*, vol. 152, pp. 122–130, 2005.

[CHO 03] CHOI P., DATTA R., "Sorption in proton-exchange membranes-an explanation of schroeder's paradox", *Journal of Electrochemical Society*, vol. 150, no. 12, pp. E601–E607, 2003.

[CHO 05] CHOI P., JALANI N.H., DATTA R., "Thermodynamics and proton transport in Nafion II. Proton diffusion mechanisms and conductivity", *Journal of the Electrochemical Society*, vol. 152, no. 3, pp. E123–E130, 2005.

[CHO 06] CHOI P., JALANI N.H., THAMPAN T.M. *et al.*, "Consideration of thermodynamic, transport, and mechanical properties in the design of polymer electrolyte membrane for higher temperature fuel cell operation", *Journal of Polymer Science B*, vol. 44, no. 16, pp. 2183–2200, 2006.

[CHO 08] CHOI J., KIM Y.H., LEE Y. *et al.*, "Numerical analysis on the performance of cooling plates in a PEFC", *Journal of Mechanical Science and Technology*, vol. 22, pp. 1417–1425, 2008.

[CHU 03] CHUPIN S., Comportement local et performances électriques d'une pile à combustible à membrane: vers un outil de diagnostic, PhD thesis, Université du Québec à Trois-Rivières, available at: http://docnum.univ-lorraine.fr/public/ INPL/2009_CHUPIN_S.pdf, 2009.

[COL 08] COLINART T., Gestion de l'eau et performances électriques d'une pile à combustible: des pores de la membrane à la cellule, PhD thesis, Institut national polytechnique de Lorraine, available at: http://docnum.univ-lorraine.fr/public/ INPL/2008_COLINART_T.pdf, 29 September 2008.

[COM 02] COMMER P., CHERSTVY A.G., SPOHR E. *et al.*, "The effect of water content on proton transport in polymer electrolyte membranes", *Fuel Cells*, vol. 2, nos 3–4, pp. 127–136, 2002.

[CON 18a] https://www.connaissancedesenergies.org/fiche-pedagogique/hydrogene-energie, accessed on May 31st, 2018.

[CON 18b] https://www.connaissancedesenergies.org/l-hydrogene-est-plus-dangereux-que-les-carburants-traditionnels, accessed on May 31st, 2018.

[COO 03] COORS W.G., "Protonic Ceramic Fuel Cells for High-Efficiency Operation with methane", *Journal of Power Sources*, vol. 118, nos 1–2, pp. 150–156, 2003.

[COS 01] COSTAMAGNA P., "Transport phenomena in polymetric membrane fuel cells", *Chemical Engineering Science*, vol. 56, no. 2, pp. 323–332, 2001.

[CUI 07] CUI S., LIU J., SELVAN M.E. *et al.*, "A molecular dynamics study of a Nafion polyelectrolyte membrane and the aqueous phase structure for Proton transport", *Journal of Physical Chemistry B*, vol. 111, no. 9, pp. 2208–2218, 2007.

[CUR 99] CURTIS C.F., BIRD R.B., "Multicomponent diffusion", *Industrial & Engineering Chemistry Research*, vol. 38, pp. 2515–2522, 1999.

[CWI 92a] CWIRKO E.H., CARBONELL R.G., "Interpretation of transport coefficients in Nafion using a parallel pore model", *Journal of Membrane Science*, vol. 67, nos 2–3, pp. 227–247, 1992.

[CWI 92b] CWIRKO E.H., CARBONELL R.G., "Ionic equilibria in ion-exchange membranes: a comparison of pore model predictions with experimental results", *Journal of Membrane Science*, vol. 67, nos 2–3, pp. 211–226, 1992.

[DIN 98] DIN X.D., MICHAELIDES E.E., "Transport processes of water and protons through micropores", *AIChE Journal*, vol. 44, no. 1, pp. 35–47, 1998.

[DJI 02] DJILALI N., LU D., "Influence of heat transfer on gas and water transport in fuel cells", *International Journal of Thermal Sciences*, vol. 41, pp. 29–40, 2002.

[DOK 06] DOKMAISRIJAN S., SPOHR E., "M D simulations of proton transport along a model Nafion surface decorated with sulfonate groups", *Journal of Molecular Liquids*, vol. 129, nos 1–2, pp. 92–100, 2006.

[DOY 03] DOYLE M., RAJENDRAN G., "Perfluorinated Membranes", in VIELSTICH W., LAMM A., GASTEGIER H.A. (eds), *Handbook of Fuel Cells, Fundamentals, Technology and Application, vol. 3: Fuel Cell technology and Application*, John Wiley & Sons, New York, 2003.

[DUT 01] DUTTA S., SHIMPALEE S., VAN ZEE J.W., "Numerical prediction of mass exchange between cathode and anode channels in a PEM fuel cell", *International Journal of Heat and Mass Transfer*, vol. 44, no. 11, pp. 2029–2042, 2001.

[EIK 98] EIKERLING M., KHARKATS Y.I., KORNYSHEV A.A. *et al.*, "Phenomenological theory of electro-osmotic effect and water management in polymer electrolyte proton conducting membranes", *Journal of the Electrochemical Society*, vol. 145, pp. 2684–2699, 1998.

[EIK 01] EIKERLING M., KORNYSHEV A.A., "Proton transfer in a single pore of a polymer electrolyte membrane", *Journal of Electroanalytical Chemistry*, vol. 502, nos 1–2, pp. 1–14, 2001.

[EIK 02] EIKERLING M., PADDISON S.J., ZAWODZINSKI T.A., "Molecular orbital calculations of proton dissociation and hydration of various acidic moieties for fuel cell polymers", *Journal of New Material for Materials for Electrochemical Systems*, vol. 5, pp. 15–25, 2002.

[ELL 01] ELLIOT J.A., HANNA S., ELLIOT A.M.S. *et al.*, "The swelling behaviour of perfluorinated ionomer membranes in ethanol/water mixtures", *Polymer*, vol. 42, no. 5, pp. 2251–2253, 2001.

[ELL 07] ELLIOTT J.A., PADDISON S.J., "Modeling of morphology and proton transport in PFSA membranes", *Physical Chemistry Chemical Physics*, vol. 9, no. 21, pp. 2602–2618, 2007.

[ERI 01] ERIKSEN J., AABERG R.J., ULLEBERG Ø. *et al.*, "System analysis of a PEMFC based stand alone power system", *1st European PEFC Forum*, Switzerland, 2001.

[FAG 05] FAGHRI A., GUO Z., "Challenges and Opportunities of thermal management issues related to fuel cell technology and modeling", *International Journal of Heat and Mass Transfer*, vol. 48, pp. 3891–3920, 2005.

[FAG 08] FAGHRI A., GUO Z., "Integration of heat pipe into fuel cell technology", *Heat Transfer Engineering*, vol. 19, pp. 232–238, 2008.

[FEL 86] FELDER R.M., ROUSSEAU R.W., *Elementary principles of chemical processes*, 2nd ed., John Wiley & Sons, New York, 1986.

[FUL 92] FULLER T.F., NEWMAN J., "Experimental determination of the transport number of water in Nafion 117 membrane", *Journal of the Electrochemical Society*, vol. 139, no. 5, pp. 1332–1336, 1992.

[FUL 93] FULLER T.F., NEWMAN J., "Water and thermal management in solid polymer electrolyte fuel cells", *Journal of the Electrochemical Society*, vol. 140, no. 5, pp. 1218–1225, 1993.

[GAL 05] GALIP H., GUVELIOGLU G.H., STENGER H.G., "Computational fluid dynamics modeling of polymer electrolyte membrane fuel cells", *Journal of Power Sources*, vol. 147, pp. 95–106, 2005.

[GAR 00] GARRETT R.H., GRISHAM C.M., LUBOCHINSKY B., *Biochimie*, De Boeck, Louvain-la-Neuve, Paris, 2000.

[GAS 03] GASTEIGER H.A., GU W., MAKHARIA R. *et al.*, "Catalyst utilization and mass transfer limitations in the polymer electrolyte fuel cells. Tutorial", *Electrochemical Society Meeting*, Orlando, FL, 2003.

[GE 03] GE S.H., YI B.L., "A mathematical model for PEMFC in different flow modes", *Journal of Power Sources*, vol. 124, pp. 1–11, 2003.

[GIE 81] GIERKE T.D., MUNN G.E., WILSON F.C., "The morphology in Nafion perfluorinated membrane products, as determined by Wide-and Small-Angle X-ray studies", *Journal of Polymer Science*, vol. 19, no. 11, pp. 1687–1704, 1981.

[GIE 82] GIERKE T.D., HSU W.Y., "The cluster-network model of ion clustering in perfuorosulfonated membranes", *ACS Symposium Series*, vol. 180, pp. 283–307, 1982.

[GLO 98] GLOAGUEN F., CONVERT P., GAMBURZEV S. *et al.*, "An evaluation of the macro-homogenous and agglomerate model for oxygen reduction in PEMFCs", *Electrochimica Acta*, vol. 43, no. 24, pp. 3767–3772, 1998.

[GOT 08] GOTTESFELD S., ZAWODZINSKI T.A., "Polymer electrolyte fuel cells", in ALKIRE R., GERISHER H., KOLB D. *et al.* (eds), *Advances in Electrochemical Science and Engineering*, vol. 5, Wiley VCH, Weinheim, 2008.

[GRO 68] GROSS J., OSTERLE J.F., "Membrane transport characteristics of ultrafine capillaries", *Journal of Chemical Physics*, vol. 49, no. 1, pp. 228–234, 1968.

[GUP 08] GUPTA R.B., *Hydrogen Fuel: Production, Transport and Storage*, CRC Press, Boca Raton, 2008.

[GUR 78] GUR Y., RAVINA I., BABCHIN A., "On the electrical double layer theory. Part II. The Poisson-Boltzmann equation including hydratation forces", *Journal of Colloid Interface Science*, vol. 64, pp. 333–341, 1978.

[GUR 98] GURAU V., LIU H., KAKAÇ S., "Two–dimensional model for proton exchange membrane fuel cells", *AIChE Journal*, vol. 44, no. 11, pp. 2410–2422, 1998.

[GUZ 90] GUZMANGARCIA A.G., PINTAURO P.N., VERBRUGGE M.W. *et al.*, "Development of a space charge transport model for ion-exchange membranes", *AIChE Journal*, vol. 36, no. 7, pp. 1061–1074, 1990.

[HAS 10] HASHMI S.M.H., Cooling Strategies for PEM FC Stacks, PhD thesis, Universität der Bundeswehr Hamburg, 2010.

[HIN 94] HINATSU J.T., MIZUHATA M., TAKENAKA H., "Water uptake of perfluorosulfonic acid membranes from liquid water and water vapour", *Journal of the Electrochemical Society*, vol. 141, pp. 1493–1498, 1994.

[HIR 94] HIRSCHENHOFER J.H., STAUFFER D.B., ENGLEMAN R.R., Fuel Cells: A Handbook (revision 3), US Department of Energy, Morgantown Energy Technology Center, DOE/METC-94/1006, January 1994.

[HIR 98] HIRSCHENHOFER J.H. *et al.*, *The Fuel Cell Handbook*, 4th ed., Parsons Corporation, Reading, 1998.

[HOR 09] HORIZON FUEL CELL TECHNOLOGIES, Renewable Energy: Science Education Manual V1, internal document, Shanghai, 2009.

[HSU 83] HSU W.Y., GIERKE T.D., "Ion transport and clustering in Nafion perfluorinated membranes", *Journal of Membrane Science*, vol. 13, pp. 307–326, 1983.

[HSU 09] HSU C.Y., WENG F.B., SU A. *et al.*, "Transient phenomenon of step switching for current or voltage in PEMFC", *Renewable Energy*, vol. 34, pp. 1979–1985, 2009.

[INC 96] INCROPERA F., DEWITT D.P., *Fundamentals of heat and mass transfer*, 4th ed., John Wiley & Sons, New York, 1996.

[INS 04] INSTITUT FRANÇAIS DU PÉTROLE (IFP), L'hydrogène: vecteur énergétique du futur?, 2004.

[ISE 99] ISE M., KREUER K.D., MAIER J., "Electroosmotic drag in polymer electrolyte membranes an electrophoretic NMR study", *Solid State Ionics*, vol. 125, pp. 213–223, 1999.

[JAN 01] JANSSEN G.J.M., "A phenomenological model of water transport in a proton exchange membrane fuel cell", *Journal of the Electrochemical Society*, vol. 148, no. 12, pp. A1313–A1323, 2001.

[KIM 95] KIM J., LEE S., SRINIVASAN S. *et al.*, "Modelling of proton exchange membrane fuel cell performance with an empirical equation", *Journal of the Electrochemical Society*, vol. 142, pp. 2670–2674, 1995.

[KJE 03] KJELSTRUP S., RØSJORDE A., "Local and total entropy production and heat and water fluxes in a one-dimensional polymer electrolyte fuel cell", *Journal of Physical Chemistry B*, vol. 109, no. 18, pp. 9020–9033, 2003.

[KOT 00] KOTER S., "The equivalent pore radius of charged membranes from electroosmotic flow", *Journal of Membrane Science*, vol. 166, no. 1, pp. 127–135, 2000.

[KOT 02] KOTER S., "Transport of simple electrolyte solutions through ion-exchange membranes the capillary model", *Journal of Membrane Science*, vol. 206, nos 1–2, pp. 201–215, 2002.

[KRE 04] KREUER K.D., PADDISON S.J., SPOHR E. *et al.*, "Transport in proton conductors for fuel cell applications: Simulation, elementary reactions and phenomenology", *Chemical Review*, vol. 104, no. 10, pp. 4637–4678, 2004.

[KUL 03] KULIKOVSKY A., "Quasi-3D modeling of water transport in polymer electrolyte fuel cells", *Journal of the Electrochemical Society*, vol. 150, no. 11, pp. A1432–A1439, 2003.

[LAL 12] LALLEMAND A., *Énergétique – Phénomènes de transfert*, Éditions Ellipses, Paris, 2012.

[LAM 93] LAMPINEN M.J., FOMINO M., "Analysis of Free Energy and Entropy changes for half-cell reactions", *Journal of the Electrochemical Society*, vol. 140, pp. 353–3546, 1993.

[LAR 03] LARMINIE J.E., DICKS A., *Fuel Cell Systems Explained*, 2nd ed., John Wiley & Sons, New York, 2003.

[LAU 01] LAURENCELLE F., CHAHINE R., HAMELIN J. *et al.*, "Characterization of a Ballard MK5-E proton exchange membrane fuel cell stack", *Fuel Cells*, no. 1, pp. 66–71, 2001.

[LEM 04] LEMAIRE T., Couplages électro-chimico-hydro-mécaniques dans les milieux argileux, PhD thesis, LEMTA – Institut national polytechnique de Lorraine, 2004.

[LI 01] LI T., WLASCHIN A., BALBUENA P.B., "Theoretical studies of proton transfer in water and model polymer electrolyte systems", *Industrial Engineering Chemical Research*, vol. 40, no. 22, pp. 4789–4800, 2001.

[LI 06] LI B.X., *Principles of Fuel Cells*, Taylor & Francis Group, New York, 2006.

[LIU 07] LIU F., LU G., WANG C.Y., "Water transport coefficient distribution though the membrane in polymer electrolyte fuel cell", *Journal of Membrane Science*, vol. 287, pp. 126–131, 2007.

[MAH 06] MAHINPEY N., JAGANNATHAN A., IDEM R., The effect of Mass transfer parameters on the modeling of PEM fuel cell, IEEE, 2006.

[MAN 00] MANN R.F., AMPHLETT J.C., HOOPER M.A.I. *et al.*, "Development and application of a generalised steady state electrochemical model for a PEM fuel cell", *Journal of Power Sources*, vol. 86, pp. 173–180, 2000.

[MAZ 03] MAZUMDER S., COLE J.V., "Rigourous 3-D mathematical modeling of PEM fuel cells", *Journal of the Electrochemical Society*, vol. 150, no. 11, pp. A1503–A1509, 2003.

[MEI 04] MEIER F., EIGENBERGER G., "Transport parameters for the modelling of water transport in ionomer membranes for pemfc fuel cells", *Electrochimica Acta*, vol. 49, no. 11, pp. 1731–1742, 2004.

[MEY 98] MEYERS J.P., Simulation and Analysis of the Direct Methanol Fuel Cell, PhD thesis, University of California Berkeley, 1998.

[MIK 01] MIKKOLA M., Experimental Studies on Polymer Electrolyte Membrane Fuel Cell Stacks, Master's thesis, Helsinki University of Technology, available at: http://www.hut.fi/Units/AES:Studies/dis/mikkola.pdf, 2001.

[MIK 07] MIKKOLA M., Studies on Limiting Factors of Polymer Electrolyte Membrane Fuel Cell Cathode Performance, PhD thesis, Helsinki University of Technology, available at: http://lib.tkk.fi/Diss/2007/isbn9789512285907/isbn9789512285907.pdf, 2007.

[MIS 13] MISE EN SERVICE DE LA CENTRALE HYBRIDE ENERTRAG: VENT, hydrogène et biogaz associés pour la production et la régulation d'énergie, press release, available at: https://www.enertrag.com/download/presse/InaugurationCI-CP-261011.pdf>, visited 27 August 2013.

[MON 06] MONROE C.W., NEWMAN J., "Onsager reciprocal relations for Maxwell-Stefan diffusion", *American Chemical Society*, vol. 45, pp. 5361–5367, 2006.

[MOT 00] MOTUPALLY S., BECKER A.J., WEIDNER J.W., "Diffusion of water in Nafion 115 membranes", *Journal of the Electrochemical Society*, vol. 147, no. 9, pp. 3171–3177, 2000.

[MUR 02] MURGIA G., PISANI L., VALENTINI M. *et al.*, "Electrochemistry and mass transport in polymer electrolyte membrane fuel cells – I. Model", *Journal of the Electrochemical Society*, vol. 149, pp. A31–A38, 2002.

[NAR 84] NAREBSKA A., KOTER S., KUJAWSKI W., "Ions and water transport across charged nafion membranes. Irreversible thermodynamics approach", *Desalination*, vol. 51, no. 1, pp. 3–17, 1984.

[NEU 99] NEUBRAND W., Modelbildung und Simulation von Elektromembranverfahren, PhD thesis, Technische Universität Stuttgart, 1999.

[NEW 91] NEWMAN J.S., *Electrochemical Systems*, Prentice Hall Inc., Englewood Cliffs, 1991.

[NGU 93a] NGUYEN T.V., WHITE R.E., "A water and heat management model for proton exchange membrane fuel cells", *Journal of Electrochemical Society*, vol. 140, no. 8, pp. 2178–2186, 1993.

[NGU 93b] NGUYEN T.V., WHITE R.E., "A water and thermal management model for proton exchange membrane fuel cells", *Journal of Electrochemical Society*, vol. 140, no. 8, 1993.

[NGU 04] NGUYEN P.T., BERNING T., DJILALI N., "Computational model of a PEM fuel cell with serpentine gas flow channels", *Journal of Power Sources*, vol. 130, pp. 149–157, 2004.

[NGU 10] NGUYEN D.A., Modélisation dynamique du cœur de pile à combustible de type PEM, PhD thesis, Institut national polytechnique de Lorraine, available at: http://docnum.univ-lorraine.fr/public/INPL/2010_NGUYEN_D_A.pdf, 2010.

[OHA 09] O'HAYRE R., SUK-WON C., COLELLA W. *et al.*, *Fuel Cell Fundamentals*, 2nd ed., Wiley & Sons, New York, 2009.

[OKA 98] OKADA T., XIE G., MEEG M., "Simulation for water management in membranes for polymer electrolyte fuel cells", *Electrochimica Acta*, vol. 43, pp. 2141–2155, 1998.

[PAD 01] PADDISON S.J., "The modeling of molecular structure and ion transport in sulfonic acid based ionomer membranes", *Journal of New Material for Materials for Electrochemical Systems*, vol. 4, no. 4, pp. 197–207, 2001.

[PAD 03] PADDISON S.J., "Proton conduction mechanisms at low degrees of hydration in sulfonic acid based polymer electrolyte membranes", *Annual Review of Materials Research*, vol. 33, pp. 289–319, 2003.

[PAU 04] PAUL R., PADDISON S.J., "The phenomena of dielectric saturation in the water domains of polymer electrolyte membranes", *Solid State Ionics*, vol. 168, nos 3–4, pp. 245–248, 2004.

[PER 08] PERAZA C., DIAZ J.G., VILLANUEVA C., "Modeling and simulation of PEM fuel cell with bond graph and 20sim", *American Control Conference*, Seattle, WA, June 2008.

[PIN 84] PINTAURO P.N., BENNION D.N., "Mass transport of electrolytes in membranes II: Determination of NaCl equilibrium and transport parameters for Nafion", *Industrial & Engineering Chemistry Fundamentals*, vol. 23, no. 2, pp. 234–243, 1984.

[PIN 89] PINTAURO P.N., VERBRUGGE M.W., "The electric potential profile in ion-exchange membrane pores", *Journal of Membrane Science*, vol. 44, nos 2–3, pp. 197–212, 1989.

[PIS 02] PISANI L., MURGIA G., VALENTINI M. *et al.*, "A new semi-empirical approach to performance curves of polymer electrolyte fuel cells", *Journal of Power Sources*, vol. 108, pp. 192–203, 2002.

[RAJ 08] RAJESHWAR K., LICHT S., MCCONNELL R., *Solar Hydrogen Generation: Toward a Renewable Energy Future*, Springer, New York, 2008.

[RAM 05] RAMOUSSE J., Transferts couplés masse-charge-chaleur dans une cellule de pile à combustible à membrane polymère, PhD thesis, Institut national polytechnique de Lorraine, available at: http://pegase.scd.inpl-nancy.fr/theses/2005_RAMOUSSE_J.pdf, 2005.

[RAM 09] RAMOUSSE J., LOTTIN O., DIDIERJEAN S. *et al.*, "Heat sources in proton exchange membrane (PEM) fuel cells", *Journal of Power Sources*, vol. 192, pp. 435–441, 2009.

[REI 09] REICHER M., Theoretische Untersuchungen zur kühlleistungssteigerung durch innovative kühlsysteme für brennstoffzellen Elektrofahrzeuge, PhD thesis, Universität Stuttgart, 2009.

[REN 97] REN X., HENDERSON W., GOTTESFELD S., "Electro-osmotic drag of water in ionmeric membranes", *Journal of the Electrochemical Society*, vol. 144, no. 9, pp. L267–L270, 1997.

[REN 01] REN X., GOTTESFELD S., "Electro-osmotic drag of water in poly (perfluorosulfonic acid) membranes", *Journal of the Electrochemical Society*, vol. 148, no. 1, pp. A87–A93, 2001.

[ROS 99] ROSSI C. *et al.*, "A new generation of MEMS based Microthrusters for Microspacecraft Application", *Proceedings of the 2nd International Conference on Integrated Micro/Nanotechnology for Space Applications*, vol. 1, pp. 201–209, 1999.

[ROS 03] ROSINI S., Capteur potentio-mètrique tout solide pour le dosage de l'hydrogène dans l'air, PhD thesis, Institut national polytechnique de Grenoble, 2003.

[ROW 01] ROWE A., LI X., "Mathematical modeling of proton exchange membrane fuel cells", *Journal Power Sources*, vol. 102, pp. 82–96, 2001.

[SAD 09] SADIQ AL BAGHDADI M.A.R., "A CFD study fo hygro thermal stresses distribution in PEM fuel cell during regular cell operation", *Renewable Energy*, vol. 34, pp. 674–682, 2009.

[SAI 04] SAISSET R., Contribution à l'étude systémique de dispositifs énergétiques à composants électrochimiques, PhD thesis, Institut national polytechnique de Toulouse, 2004.

[SAI 06] SAISSET R., FONTES G., TURPIN C. *et al.*, "Bond Graph model of a PEM fuel cell", *Journal of Power Sources*, vol. 156, pp. 100–107, 2006.

[SEE 05] SEELIGER D., HARTNIG C., SPOHR E., "Aqueous pore structure and proton dynamics in solvated Nafion membranes", *Electrochimica Acta*, vol. 50, no. 21, pp. 4234–4240, 2005.

[SER 07] SERICAN M.F., YESILYURT S., "Transient analysis of proton electrolyte membrane fuel cells (PEMFC) at start-up and failure", *Fuel Cells*, vol. 7, no. 2, pp. 118–127, 2007.

[SHA 03] SHAH R.K., "Heat exchangers for fuel cell systems", *The 4th International Conference on Compact Heat Exchangers and Enhancement Technology for the Process Industries*, Crete, Greece, 2003.

[SHI 06] SHI Z., WANG X., ZHANG Z., "Comparison of two-dimensional PEM fuel cell modeling using COMSOL multiphysics", *Proceedings of the COMSOL Users Conference*, Boston, 2006.

[SIE 03] SIEGEL N.P., ELLIS M.W., NELSON D.J. *et al.*, "Single domain PEMFC model based on agglomerate catalyst geometry", *Journal of Power Sources*, vol. 115, pp. 81–89, 2003.

[SIE 08] SIEGEL C., "Review of computational heat and mass transfer modeling in polymer electrolyte membrane (PEM) fuel cells", *Energy*, vol. 33, pp. 1331–1352, 2008.

[SIN 99] SINGH D., LU D.M., DJILALI N., "A two-dimensional analysis of mass transport in proton exchange membrane fuel cells", *International Journal of Engineering Science*, vol. 37, pp. 431–452, 1999.

[SME 00] SMEDLEY S., "A Regenerative Zinc – Air Fuel Cell for industrial and specialty vehicles", *Fifteenth Annual Battery Conference on Applications and Advances*, Long Beach, January 2000.

[SON 07] SONG C., TANG Y., ZHANG J.L. *et al.*, "PEM fuel cell reaction kinetics in the temperature range of 23–120 °C", *Electrochemica Acta*, vol. 52, pp. 2552–2561, 2007.

[SOR 02] SORENSENA T.S., KJELSTRUP S., "A simple Maxwell-Wagner-Butler-Volmer approach to the impedance of the hydrogen electrode in a nafion fuel cell", *Journal of Colloid and Interface Science,* vol. 248, no. 2, pp. 355–375, 2002.

[SPI 07] SPIEGEL C., *Designing and building fuel cells*, McGraw Hill, New York, 2007.

[SPO 02] SPOHR E., COMMER P., KORNYSHEV A.A., "Enhancing proton mobility in polymer electrolyte membranes: lessons from molecular dynamics simulations", *Journal of Physical Chemistry B*, vol. 106, no. 41, pp. 10560–10569, 2002.

[SPR 91] SPRINGER T.E., ZAWODZINSKI T.A., GOTTESFELD S., "Polymer electrolyte fuel cell model", *Journal of the Electrochemical Society*, vol. 138, no. 8, pp. 2334–2342, 1991.

[SPR 93] SPRINGER T.E., WILSON M.S., GOTTESFELD S., "Modeling and experimental diagnostics in polymer electrolyte fuel cells", *Journal of Electrochemical Society*, vol. 140, no. 12, pp. 3513–3526, 1993.

[SRI 06] SRINIVASSAN S., *Fuel Cells from fundamentals to applications*, Springer, Berlin, 2006.

[TAN 97] TANDON R., PINTAURO P.N., "Divalent/monovalent cation uptake selectivity in a Nafion cation exchange membrane: Experimental and modeling studies", *Journal of Membrane Science*, vol. 136, nos 1–2, pp. 207–219, 1997.

[THI 97] THIRUMALAI D., WHITE R.E., "Mathematical modelling of proton exchange membrane fuel cell stacks", *Journal of the Electrochemical Society*, vol. 144, pp. 1717–1723, 1997.

[VAL 06] VALLERES C., WINKELMANN D., ROIZARD D. *et al.*, "On Schröder's Paradox", *Journal of Membrane Science*, vol. 278, nos 1–2, pp. 357–364, 2006.

[VAN 04] VANDERSTEEN J.D.J., KENNEY B., PHAROAH J.G. *et al.*, "Mathematical modelling of the transport phenomena and the chemical/electrochemical reactions in solid oxide fuel cells: a review", *Proceedings of the Canadian Hydrogen and Fuel Cells Conference*, Toronto, September 2004.

[VER 88] VERBRUGGE M.W., HILL R.H., "Experimental and theoretical investigation of perfluorosulfonic acid membrane equilibrated with aqueous sulfuric acid solutions", *Journal of Physical Chemistry*, vol. 92, no. 23, pp. 6778–6783, 1988.

[WAG 03] WAGNER H.A., "Space – shuttle fuel cell", in VIELSTICH W., LAMM A., GASTEIGER H.A. (eds), *Handbook of Fuel Cells, Fundamentals technology and application*, vol. 4., John Wiley & Sons, New York, 2003.

[WAN 04] WANG C.Y., "Fundamentals models for fuel cell engineering", *Chemical Reviews*, vol. 104, pp. 4727–4766, 2004.

[WAN 06a] WANG Y., WANG C.Y., "Dynamics of polymer electrolyte fuel cells undergoing load changes", *Electrochimica Acta*, vol. 51, pp. 3924–3933, 2006.

[WAN 06b] WANG Y., WANG C.Y., NONISOTHERMAL A., "Two-phase model for polymer electrolyte fuel cells", *Journal of the Electrochemical Society*, vol. 153, no. 6, pp. A1193–A1200, 2006.

[WAT 03] WATARI T., WANG H., KUWAHARA K. *et al.*, "Water vapor sorption and diffusion properties of sulfonated polyimide membranes", *Journal of Membrane Science*, vol. 219, pp. 137–147, 2003.

[WEB 04a] WEBER A.Z., NEWMAN J., "Modeling transport in polymer electrolyte fuel cells", *Chemical Reviews*, vol. 104, pp. 4679–4726, 2004.

[WEB 04b] WEBER A.Z., NEWMAN J., "Transport in polymer electrolyte membranes – II. Mathematical model", *Journal of the Electrochemical Society*, vol. 151, no. 2, pp. A311–A325, 2004.

[WEB 06] WEBER A.Z., NEWMAN J., "Coupled thermal and water management in polymer electrolyte fuel cells", *Journal of the Electrochemical Society*, vol. 153, no. 12, pp. A2205–A2214, 2006.

[WES 96] WEST A.C., FULLER T.F., "Influence of rib spacing in proton-exchange membrane electrodes assemblies", *Journal of Applied Electrochemistry*, vol. 26, no. 6, pp. 557–565, 1996.

[WES 06] WESCOTT J.T., QI Y., SUBRAMANIAN L. *et al.*, "Mesoscale simulation of morphology in hydrated perfluorosulfonic acid membranes", *Journal of Chemical Physics*, vol. 124, no. 13, p. 134702, 2006.

[WÖH 00] WÖHR M., Instationäres, Thermodynamisches Verhalten der Polymermembrane-Brennstoffzelle, PhD thesis, Dusseldorf, 2000.

[WÖH 98] WÖHR M., BOLWIN K., SCHNUMBERGER W. *et al.*, "Dynamic modeling and simulation of a polymer membrane fuel cell including mass transport limitation", *International Journal of Hydrogen Energy*, vol. 23, pp. 213–218, 1998.

[WU 07] WU H., BERG P., LI X., "Non-isothermal transient modeling of water transport inPEM fuel cells", *Journal of Power Sources*, vol. 165, pp. 232–243, 2007.

[XU 07] XU C., ZHAO T.S., "A new flow field design for polymer electrolyte based fuel cells", *Electrochemistry Communications*, vol. 9, pp. 497–503, 2007.

[YAN 00] YANG Y., PINTAURO P.N., "Multicomponent space-charge transport model for ion-exchange membranes", *AIChE Journal*, vol. 46, no. 6, pp. 1177–1190, 2000.

[YAN 04a] YAN W.M., CHEN F., WU H.Y. *et al.*, "Analysis of thermal and water management with temperature-dependent diffusion effects in membrane of proton exchange membrane fuel cells", *Journal Power Sources*, vol. 129, pp. 127–137, 2004.

[YAN 04b] YANG Y., PINTAURO P.N., "Multicomponent space-charge transport model for ion-exchange membranes with variable pore properties", *Industrial Engineering Chemical Research*, vol. 43, no. 12, pp. 2957–2965, 2004.

[YE 07] YE X., WANG C.Y., "Measurement of water transport properties through membrane-electrode assemblies", *Journal of the Electrochemical Society*, vol. 154, no. 7, pp. B676–B682, 2007.

[YEO 77] YEO S.C., EISENBERG A., "Physical properties and supermolecular structure of perfluorinated ion-containing (Nafion) polymers", *Journal of Applied Polymer Science*, vol. 21, no. 4, pp. 875–898, 1977.

[YI 99] YI J.S., NGUYEN T.V., "Multicomponent transport in porous electrodes of proton exchange membrane fuel cells using the interdigitated gas distributors", *Journal of the Electrochemical Society*, vol. 146, pp. 38–45, 1999.

[YU 09] YU S.H., SOHN S., NAM J.H. *et al.*, "Numerical study to examine the performance of multi-pass serpentine flow fields for cooling plates in polymer electrolyte membrane fuel cells", *Journal of Power Sources*, vol. 194, pp. 697–703, 2009.

[ZAW 93a] ZAWODZINSKI T.A., DEROUIN C., RADZINSKI S. *et al.*, "Water uptake by and transport through Nafion 117 membranes", *Journal of Electrochemical Society*, vol. 140, pp. 1041–1047, 1993.

[ZAW 93b] ZAWODZINSKI T.A., SPRINGER T.E., URIBE F. *et al.*, "Characterization of polymer electrolytes for fuel cell applications", *Solid State Ionics*, vol. 60, pp. 199–211, 1993.

[ZAW 95] ZAWODZINSKI T.A., DAVEY J., VALERIO J. *et al.*, "The water dependence of electroosmotic drag in proton-conducting polymer electrolytes", *Electrochimica Acta*, vol. 40, no. 3, pp. 297–302, 1995.

[ZHA 06] ZHANG J., XIE Z., ZHANG J. *et al.*, "High temperature PEM fuel cells", *Journal of Power Sources*, vol. 160, pp. 872–891, 2006.

[ZIE 05] ZIEGLER C., Modeling and simulation of the dynamic behavior of portable proton exchange membrane fuel cells, PhD thesis, Fraunhofer Institut für Solare Energiesysteme Freiburg im Breisgau, 2005.

Index

Printed in the United States
By Bookmasters